Felix Fritzen

Microstructural modeling and computational homogenization of the physically linear and nonlinear constitutive behavior of micro-heterogeneous materials

Schriftenreihe
Kontinuumsmechanik im Maschinenbau
Band 1

Karlsruher Institut für Technologie
Institut für Technische Mechanik
Bereich Kontinuumsmechanik

Hrsg. Prof. Dr.-Ing.habil. Thomas Böhlke

Microstructural modeling and computational homogenization of the physically linear and nonlinear constitutive behavior of micro-heterogeneous materials

by
Felix Fritzen

KIT Scientific Publishing

Dissertation, Karlsruher Institut für Technologie
Fakultät für Maschinenbau
Tag der mündlichen Prüfung: 12. Mai 2011

Impressum

Karlsruher Institut für Technologie (KIT)
KIT Scientific Publishing
Straße am Forum 2
D-76131 Karlsruhe
www.ksp.kit.edu

KIT – Universität des Landes Baden-Württemberg und nationales
Forschungszentrum in der Helmholtz-Gemeinschaft

KIT Scientific Publishing 2011
Print on Demand

ISSN: 2192-693X
ISBN: 978-3-86644-699-1

Microstructural modeling and computational homogenization of the physically linear and nonlinear constitutive behavior of micro-heterogeneous materials

Zur Erlangung des akademischen Grades

Doktor der Ingenieurwissenschaften

der Fakultät für Maschinenbau

Karlsruher Institut für Technologie (KIT)

genehmigte

Dissertation

von

Dipl.-Ing. Dipl.-Math.techn. Felix Fritzen

Tag der mündlichen Prüfung	12.05.2011
Hauptreferent	Prof. Dr.-Ing.habil. Thomas Böhlke
Korreferenten	Prof. Dr.-Ing.habil. Samuel Forest
	Prof. Dr.-Ir. M.G.D. Geers

Acknowledgement

Throughout the entire creation of the thesis, numerous people have either explicitly or implicitly helped me to attain the results in the present form. Hereby, I would like to express my gratitude to my colleagues at the Chair for Continuum Mechanics. They have created a friendly ambience and we have shared endless discussions. Particular thanks go to Prof. Thomas Böhlke for the provided freedom with respect to the choice of the pursued research. The given confidence has helped me to explore interesting research directions based on personal preference.

Further, I highly acknowledge the atmosphere created by the people at the *Centre des Matériaux* (Paris) during my research stay in the summer of 2010. Particular thanks go to Prof. Samuel Forest for the assignment of the research objective pursued in that period, as well as for the countless stimulating discussions we shared. The time there not only helped to improve my French but also provided me with a stimulating surrounding that eased the editing of parts of the thesis considerably. Financial support for the stay was provided through means of the *Exzellenzinitiative* via the KHYS (Karlsruhe House of Young Scientists) and was highly appreciated.

Additional thanks go to Prof. Marc Geers and Prof. Samuel Forest who kindly reviewed the work. It was a pleasure that both of them were able to attend the presentation and the subsequent examination, despite their many professional obligations.

Last but not least I would like to express my thanks to all the people outside of the scientific scope, namely my friends and my family. They offered me great support and showed lots of confidence, which helped to stay motivated during the past four years.

Contents

1 Introduction **5**

2 Basic equations **9**

2.1 Kinematics 9

2.1.1 General kinematic relations 9

2.1.2 Deformation measures 11

2.2 Constitutive Modeling 12

2.2.1 Mechanical field variables and balance of momentum 12

2.2.2 Thermodynamic field variables and energy balance 14

2.2.3 Generalized Standard Materials 17

3 Real microstructures and model representations **19**

3.1 Motivation 19

3.2 Statistical properties of microstructures 22

3.2.1 Microstructural n-point statistics 23

3.2.2 Topology dependent statistical descriptions 24

3.2.2.1 Materials with convex granular structure 24

3.2.2.2 Porous materials 25

3.3 Artificial microstructures based on random Voronoi tessellations 28

3.3.1 Representation of crystal aggregates 28

3.3.1.1 Modification of the generator points 30

3.3.1.2 Consideration of anisotropic grain morphology 32

3.3.1.3 Periodic spatial discretization of crystal aggregates 34

3.3.1.4 Example meshes 38

3.3.2 Particle reinforced composites 41

3.3.2.1 Mesh generation algorithm for particulate materials 42

3.3.2.2 Example meshes 44

3.4 Porous materials 46

3.4.1 Microstructure generation 46

3.4.2 Mesh generation for porous materials 49

4 Homogenization methods **51**

4.1 Thermo-mechanical two-scale problems 51

4.1.1 Boundary conditions for homogenization methods 55

4.2 Linear homogenization problems 57
4.3 Semi-analytical methods 61
4.4 Computational homogenization 65
4.4.1 Linear computational homogenization 65
4.4.2 Approaches to computational nonlinear homogenization 66

5 Non-uniform transformation field analysis 71
5.1 Introduction 71
5.2 Relation to order reduction methods 72
5.2.1 Solution to the microscopic problems 74
5.2.2 Thermodynamic driving forces 75
5.3 Analysis of the system matrices 78
5.4 Evolution of mode activity 80
5.4.1 Existing approaches 80
5.4.2 Verification and falsification 82
5.4.3 The uncoupled model 84
5.4.4 The coupled model 85
5.4.5 A modified coupled model 86
5.5 Mode identification 88
5.5.1 Numerical Experiments 88
5.5.2 Karhunen-Loève decomposition 89
5.5.3 Alternative kinematic mode identification procedure 91
5.5.4 A thermodynamically motivated approach 92
5.6 Numerical implementation of the non-uniform transformation field analysis 95
5.6.1 Data processing steps 95
5.6.2 Finite element operators 97
5.6.3 Implicit time integration for rate-independent materials 99
5.6.4 Algorithmic tangent operators for rate-independent materials 102

6 Computational results 105
6.1 Homogenization of crystalline aggregates 105
6.1.1 Elastic properties of cubic crystal aggregates 105
6.1.2 Influence of the grain morphology on elasto-visco-plastic properties 108
6.2 Thermo-elastic properties of metal ceramic composites 111
6.3 Inelastic homogenization of porous metals 118
6.3.1 Constitutive assumptions 118
6.3.2 Boundary conditions 118
6.3.3 Statistical properties of the microstructures 120
6.3.4 Mesh density convergence study 121
6.3.5 Asymptotic results 123
6.3.6 Representativity of the results 123
6.3.7 Local plastic strain fields 127
6.4 Interpretation and discussion 128
6.4.1 Comparison with analytical models 128

6.4.2 Identification of a modified GTN model 130
6.5 Non-uniform transformation field analysis of metal matrix composites . . . 132
6.5.1 Considered microstructures 132
6.5.2 Comparison to full-field simulations on integration point level . . . 135
6.5.3 Assessment of different mode identification strategies 141
6.5.4 Structural applications 147
6.5.4.1 Tension experiment on a dog-bone specimen 147
6.5.4.2 Indentation test 147

7 Summary and conclusions 151

A Notation 155
A.1 General tensorial notation 155
A.2 Special tensorial basis 157
A.3 Vector-matrix representation of symmetric tensors 158
A.4 List of abbreviations 159

Zusammenfassung

Die meisten im Ingenieurbau eingesetzten Materialien werden auf der Strukturebene als homogene Werkstoffe interpretiert, obwohl die Mikrostruktur der Werkstoffe zum Teil erhebliche Heterogenitäten aufweist. Die makroskopisch homogenen Eigenschaften hängen häufig stark von der Mikrostuktur ab. In dieser Arbeit werden Methoden zur approximativen Beschreibung der Mikrostrukturgeometrie durch Modellstrukturen vorgestellt. Diese Modellstrukturen werden durch Zufallszahlen parametrisiert, so dass die statistischen Eigenschaften der virtuellen Materialien steuerbar sind. Maße zur statistischen Charakterisierung werden eingeführt, um einen Vergleich zwischen realen und artifiziellen Geometrien sowie zwischen verschiedenen Realisierungen der Modellmikrostrukturen zu ermöglichen.

Zur Verbesserung der numerischen Effizienz in der nachgeschalteten numerischen Homogenisierung erfolgt die Generierung der Strukturen ausschließlich periodisch. Detaillierte Algorithmen zur periodischen räumlichen Diskretisierung der geometrischen Modelle werden vorgestellt. Diese dienen der computergestützten thermo-mechanischen Homogenisierung der Eigenschaften unter Verwendung der Finiten Elemente Methode. Eine Darstellung der thermo-mechanisch gekoppelten Bilanzgleichung für Zwei-Skalenprobleme wird hergeleitet. Auf Basis der präsentierten Homogenisierungsmethoden und Diskretisierungsmethoden werden die linear elastischen Eigenschaften polykristalliner Metalle mit isotroper Gitterorientierungsverteilung in einer Monte-Carlo-Studie untersucht. Der Einfluss morphologisch und/oder kristallographisch anisotroper Polykristalle auf das elasto-visko-plastische makroskopische Verhalten wird analysiert. Weiterhin werden die thermo-elastischen Eigenschaften von Metall-Matrix-Verbundwerkstoffen mit Partikelverstärkung numerisch untersucht, wobei auch der Einfluss der Homogenität der Einschlussgeometrie berücksichtigt wird.

Für die nicht-lineare Homogenisierung werden zwei Ansätze verfolgt: Die *Nonuniform Transformation Field Analysis* (NTFA) und die numerische Parameteridentifikation. Bei der NTFA handelt es sich um ein Ordnungsreduktionsverfahren mit mikromechanischer Motivation, das durch eine geeignete Parametrisierung der plastischen Dehnungen das hochdimensionale, nicht-lineare Problem auf einen niedrigdimensionalen Lösungsraum reduziert. Die Methode wird im Detail untersucht, reformuliert und erweitert. Eine exakte Analyse der auftretenden Koeffizientenmatrizen erfolgt, und die numerische Implementierung wird detailliert dargestellt. Zur Verifikation der Methode werden zahlreiche Beispiele von Metall-Matrix-Verbundwerkstoffen numerisch untersucht. Die numerische Effizienz der NTFA wird diskutiert und bewiesen.

Im Rahmen der numerischen Parameteridentifikation werden anhand von Monte-Carlo-Simulationen makroskopische Fließkurven für poröse Metalle untersucht. Durch die numerischen Ergebnisse ist eine Modifikation bestehender Ansätze möglich, die lediglich einen weiteren Parameter benötigt. Der Vergleich der erhaltenen Fließkurven führt auf eine gute Übereinstimmung mit den Simulationsergebnissen für alle untersuchten Mikrostrukturen.

Summary

Most engineering materials are considered homogeneous on a structural level although the microstructure of the materials may show a pronounced heterogeneity. The macroscopically homogeneous constitutive behavior often depends heavily on this underlying microstructure. Therefore, methods for the approximative description of the real microstructural geometry in terms of model microstructures are presented. The artificial materials are parametrized by random variables such that the statistical properties of the virtual materials can be controlled. In order to compare real and artificial materials, as well as different statistical realizations of the artificial structures, measures for the statistical characterization are derived.

Periodic microstructures are generated in order to attain improved numerical efficiency during the numerical homogenization process. Detailed algorithms for the periodic spatial discretization of the geometric models are presented. These discretizations are used in the computer aided thermo-mechanical homogenization based on the finite element method. A representation of the strongly coupled thermo-mechanical balance equations are derived for a two-scale problem. The presented homogenization and discretization methods are used to evaluate the linear elastic properties of polycrystalline metals with isotropic crystallite orientation distribution in a Monte Carlo type study. The influence of morphological and/or crystallographic texture of polycrystals onto the elasto-visco-plastic macroscopic behavior is analyzed. Additionally, the thermo-elastic material properties of metal matrix composites with particle reinforcement are computationally investigated. In this investigation the regularity of the particle shape is also considered.

Further, two distinct nonlinear homogenization methods are examined: The *Nonuniform Transformation Field Analysis* (NTFA) and the direct computational parameter identification method. The NTFA is an order reduction method with micro-mechanical motivation. Based on a suitable parametrization of the inelastic strains the high-dimensional nonlinear problem is reduced to a low-dimensional subspace. The method is examined in detail. Reformulations of the underlying equations and extensions are presented. Additionally, the occurring coefficient matrices are investigated and the numerical implementation is covered. A verification of the method is pursued in terms of a variety of numerical examples for metal ceramic composites. The numerical efficiency of the method is discussed.

In the context of the numerical parameter identification, Monte Carlo type simulations on periodic porous aggregates are performed in order to obtain the macroscopic yield surface of the material. The numerical results of the latter allow for a modification of existing approaches in terms of a single additional parameter. The comparison of the thereby generated yield curves leads to an excellent agreement with the simulated ones for all examined microstructures.

Chapter 1

Introduction

In engineering applications real materials are usually considered on a structural (or macroscopic) level at which certain material properties such as the Young's modulus E are observed. Examples for materials commonly used in mechanical applications are sheet metal, polycrystalline metals (bulk material) or classical composites containing particulate or fibrous reinforcements. While these materials appear homogeneous from a structural point of view, observations at other length scales show a different picture. The underlying microstructure is usually a geometrically complex ensemble containing a variety of different micro-constituents. These exhibit different material properties and often local preferred directions are found, i.e. physical anisotropy of the material can locally be found. But not only the physical properties influence the effective behavior of the material. Additionally the local shape and topology of the material has an influence on the macroscopic properties due to local interactions of the phases implied by the governing balance equations (Nemat-Nasser and Hori, 1999; Torquato, 2002).

An increasing demand for high quality predictions of the thermo-mechanical response of microheterogeneous materials originates from industrial applications. Here the improved weight to strength ratio of the materials is often the driving factor. An understanding of the local mechanisms and the effects of microstructural changes can help to improve the efficiency of materials. This concerns not only classical composites such as fiber reinforced thermo-plastics or metal ceramic composites, but also biomaterials which are often highly heterogeneous. In order to allow for realistic predictions of the structural behavior an understanding of the linear elastic and thermal properties is often not sufficient. More precisely the consideration of plasticity or damage is often required.

The presented work is devoted to methods for the numerical prediction of the overall linear and nonlinear thermo-mechanical properties of materials with microstructure. The basic mechanical theory of the underlying problems is briefly revised in chapter 2. Due to the complexity of experiments on the microscopic scale the concept of model microstructures is motivated in chapter 3. In order to characterize artificial materials and to compare them to real microstructures, some elements of the statistical description of microstructures are introduced in section 3.2 (see, e.g., Ohser and Mücklich, 2000; Torquato, 2002, for more details). Several important classes of artificial materials and details on their spatial discretization for the use with the finite element method are discussed in sections 3.3, 3.4. First, polyhedral granular structures based on the Voronoi tessellation (e.g. Aurenhammer, 1991) are discussed. The resulting microstructures resemble polycrystals. In order to influence the shape of the grains different modifications of the generator points and anisotropic affine transformations of the microstructures are investigated. Special emphasis is on the periodicity of the microstructures and of the finite element discretization. The latter has strong implications on the statistical representativity and on the numerical properties. Based on the granular microstructures a method for the generation of periodic polyhedral particles embedded into a matrix material is derived. Further, porous materials based on an assemblage of hard spheres are investigated. For the granular and the porous microstructures the periodicity of the geometry and the spatial discretization are also enforced.

In chapter 4 the definition of the considered class of two-scale problems is presented. The underlying assumptions characterizing the latter are summarized. Formulations of the balance of momentum and the balance of energy on both scales are given. The microscopic boundary conditions coupling the two scales are briefly discussed. The presented framework is used to derive a formulation for the homogenization of the linear elastic, thermal and thermo-elastic properties. A general classification of homogenization techniques and a concise introduction of semi-analytical homogenization methods is given in section 4.3. The computational realization of the homogenization of the linear thermo-elastic properties is then described in section 4.4.1.

While linear homogenization problems are well understood, many problems can arise from the physical nonlinearity of one or several micro-constituents. In particular these materials show a behavior that has an intriguing dependency on the loading path. Problems resulting from this fact are outlined in section 4.4.2. Additionally, some computational approaches to the topic are briefly introduced and the use of hybrid schemes in terms of the transformation field analysis (TFA) (Dvorak and Benveniste, 1992; Dvorak et al., 1994a) and the derived nonuniform transformation field analysis (NTFA) are shortly addressed.

The NTFA (Michel and Suquet, 2003, 2004; Fritzen and Böhlke, 2010b) is extensively discussed in chapter 5. This method is an order reduction technique specifically designed for homogenization problems. It reduces the inelastic strain field found in a microstructured material to a finite dimensional but spatially heterogeneous basis of nonuniform transformation strains. A sound micro-mechanical derivation for the construction of the homogenized material model is presented. Theoretical aspects concerning the micro-mechanically derived system matrices of the homogenized material response are formulated in a thermo-mechanical framework in section 5.3. Existing approaches for an approximation of the homogenized rate potential determining the evolution of the inelastic variables are analyzed in section 5.4.2. These investigations confirm the coupled model of Michel and Suquet (2003). A detailed review on existing mode identification strategies used by Roussette et al. (2009) and Fritzen and Böhlke (2010b) is given in section 5.5. Incorporating the structure of the thermodynamic driving forces allows for a new mode identification procedure involving micro-mechanical implications.

The numerical implementation of the NTFA is described in full detail in section 5.6. Aspects related to the finite element implementation of some of the relevant field operators are provided. Moreover, the local constitutive equations of the homogenized material model are integrated using an implicit Euler time integration procedure. The Jacobian of the Newton scheme used in the time integration procedure and the consistent tangent stiffness operator of the composite are derived.

In chapter 6 the methods presented in the previous chapters are applied to different materials to outline their efficiency. First, the effective linear elastic properties of polycrystalline aggregates consisting of single crystals with cubic elastic symmetry are determined using computational homogenization in section 6.1.1. A large number of statistical realizations of periodic discretizations of Voronoi tessellations is considered. The results of the numerical homogenization are compared to bounds of odd order derived by Dederichs and Zeller (1973); Zeller and Dederichs (1973). Concerning the influence of the grain shape of elasto-visco-plastic copper polycrystals an anisotropic geometry is used in section 6.1.2.

The thermo-elastic properties of particulate metal matrix composites are determined in a Monte Carlo type study in section 6.2. Periodic finite element discretizations of materials with particle volume fractions ranging from 10 to 80% are considered to highlight the flexibility of the proposed methodology. Aspects of the regularity of the grain shape are analyzed to find the influence of higher statistical moments onto the effective elastic response. The computational results are compared to some analytical estimates.

In section 6.5 different examples of the NTFA are presented. In a first step the accuracy and the numerical performance of the method is verified for particulate metal ceramic composites (18.2% and 20% particles) by comparing the homogenized stress response with full-field finite element simulations. Different particle morphologies are examined, i.e. the aspect ratios of the particles has been varied from oblate to elongated. The load partitioning between the phases is also computed in order to examine the ability to predict not only the global stress response, but also the phase averages accurately. The robustness of the method is highlighted in terms of a comparison of different mode identification strategies for a composite with 40% particles. Finally, the homogenized material model containing only a handful of coefficients is used in macroscopic problems including contact with success.

Chapter 2

Basic equations

2.1 Kinematics

2.1.1 General kinematic relations

Let $\mathscr{B}(t)$ a general simply connected body at a given time t. Attention is confined to the observation of processes during the time interval $T = [t_0, \hat{t}]$ $(\hat{t} > t_0)$. The position of the body at the initial time t_0 is taken as the referential stress-free (relaxed) placement of the body. In the following this particular configuration is referred to as $\mathscr{B}_0$ and it serves to identify material points (or particles) $\boldsymbol{X} \in \mathscr{B}_0 \subset \mathbb{R}^3$ in analogy to Truesdell and Noll (1965). During the deformation process the spatial position of the particles of the body may change. The movement of a particle $\boldsymbol{X}$ is described in terms of the motion $\boldsymbol{\chi}(\boldsymbol{X}, t)$ (Fig. 2.1).

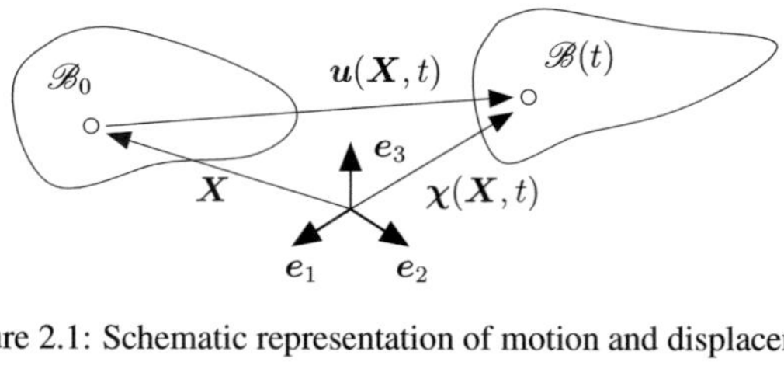

Figure 2.1: Schematic representation of motion and displacement

The motion of the body is assumed to be invertible, i.e. a point found at place $\boldsymbol{x}$ in the current placement can uniquely be identified in the referential placement and vice versa. Requiring invertibility of the motion is identical to precluding self-penetration and fracture. For convenience the spatial position of the particle $\boldsymbol{X}$ at time t is written as $\boldsymbol{x}$, where the arguments are omitted if not explicitly required. While $\boldsymbol{\chi}(\boldsymbol{X}, t)$ is a description of the absolute position of the body, one is usually interested in the movement of points relative to their referential placement. Therefore, the displacement $\boldsymbol{u}(\boldsymbol{X}, t)$ is introduced. The latter is related to the motion via

$$\boldsymbol{u}(\boldsymbol{X}, t) = \boldsymbol{\chi}(\boldsymbol{X}, t) - \boldsymbol{X}. \tag{2.1}$$

In order to describe the time-dependency of a Lagrangian field $\phi(\boldsymbol{X}, t)$, the material time derivative is defined by

$$\dot{\phi}(\boldsymbol{X}, t) = \frac{\partial \phi(\boldsymbol{X}, t)}{\partial t}. \tag{2.2}$$

Then the velocity and the acceleration of a particle are given by

$$\boldsymbol{v}(\boldsymbol{X}, t) = \dot{\boldsymbol{x}} = \frac{\partial \boldsymbol{\chi}(\boldsymbol{X}, t)}{\partial t} = \frac{\partial \boldsymbol{u}(\boldsymbol{X}, t)}{\partial t}, \tag{2.3}$$

$$\boldsymbol{a}(\boldsymbol{X}, t) = \ddot{\boldsymbol{x}} = \frac{\partial^2 \boldsymbol{\chi}(\boldsymbol{X}, t)}{\partial t^2} = \frac{\partial^2 \boldsymbol{u}(\boldsymbol{X}, t)}{\partial t^2}, \tag{2.4}$$

respectively. The derivative of ϕ with respect to the spatial (i.e. current) position is

$$\operatorname{grad}(\phi_{\mathrm{E}}(\boldsymbol{x}, t)) = \frac{\partial \phi_{\mathrm{E}}(\boldsymbol{x}, t)}{\partial \boldsymbol{x}} = \frac{\partial \phi(\boldsymbol{\chi}^{-1}(\boldsymbol{x}, t), t)}{\partial \boldsymbol{x}}, \tag{2.5}$$

where ϕ_{E} is the Eulerian representation of ϕ defined by:

$$\phi_{\mathrm{E}}(\boldsymbol{x}, t) = \phi(\boldsymbol{\chi}^{-1}(\boldsymbol{x}, t), t). \tag{2.6}$$

The derivative with respect to material (i.e. referential) coordinates is defined by

$$\operatorname{grad}_{\boldsymbol{X}}(\phi(\boldsymbol{X}, t)) = \frac{\partial \phi(\boldsymbol{X}, t)}{\partial \boldsymbol{X}}. \tag{2.7}$$

Accordingly, the divergence with respect to material and spatial coordinates is

$$\operatorname{div}_{\boldsymbol{X}}(\phi(\boldsymbol{X}, t)) = \operatorname{tr}(\operatorname{grad}_{\boldsymbol{X}}(\phi(\boldsymbol{X}, t))), \quad \operatorname{div}(\phi_{\mathrm{E}}(\boldsymbol{x}, t)) = \operatorname{tr}(\operatorname{grad}(\phi_{\mathrm{E}}(\boldsymbol{x}, t))). \tag{2.8}$$

With these definitions the material time derivative of an Eulerian field $\psi(\boldsymbol{x}, t)$ is

$$\dot{\psi}(\boldsymbol{x}, t) = \frac{\partial \psi(\boldsymbol{x}(\boldsymbol{X}, t), t)}{\partial t} = \frac{\partial \psi(\boldsymbol{x}, t)}{\partial \boldsymbol{x}} \cdot \boldsymbol{v}(\boldsymbol{X}, t) + \frac{\partial \psi(\boldsymbol{x}, t)}{\partial t}. \tag{2.9}$$

In the following it is assumed that the motion is twice continuously differentiable with respect to time at almost all time points and twice continuously differentiable with respect to the spatial position almost everywhere, where the term *almost everywhere* refers to the Borel Lebesgue measure in $\mathbb{R}$ and $\mathbb{R}^3$, respectively.

2.1.2 Deformation measures

In order to describe the behavior of materials with respect to a given deformation it is necessary to find a suitable quantity describing the latter. When considering two distinct material points $\boldsymbol{X}$ and $\boldsymbol{Y} = \boldsymbol{X} + \mathrm{d}\boldsymbol{X}$ ($\mathrm{d}\boldsymbol{X} \neq \boldsymbol{0}$) the motion of $\boldsymbol{Y}$ can be constructed by expanding the Taylor series

$$\chi(\boldsymbol{Y},t) = \chi(\boldsymbol{X},t) + \mathrm{grad}_{\boldsymbol{X}}(\chi(\boldsymbol{X},t))\,\mathrm{d}\boldsymbol{X} + \mathcal{O}\left(|\,\mathrm{d}\boldsymbol{X}|^2\right). \tag{2.10}$$

In the limit case of an infinitesimal line element $\mathrm{d}\boldsymbol{X}$ connecting the material points, the last term in (2.10) vanishes and

$$\lim_{||\,\mathrm{d}\boldsymbol{X}||\to\boldsymbol{0}} (\chi(\boldsymbol{Y},t) - \chi(\boldsymbol{X},t)) = \boldsymbol{y} - \boldsymbol{x} = \mathrm{d}\boldsymbol{x} = \mathrm{grad}_{\boldsymbol{X}}(\chi(\boldsymbol{X},t))\,\mathrm{d}\boldsymbol{X} \tag{2.11}$$

is obtained for arbitrary infinitesimal line elements $\mathrm{d}\boldsymbol{X}$. The latter defines the gradient of the motion

$$\boldsymbol{F}(\boldsymbol{X},t) = \mathrm{grad}_{\boldsymbol{X}}(\chi(\boldsymbol{X},t)) = \frac{\partial \boldsymbol{x}(\boldsymbol{X},t)}{\partial \boldsymbol{X}} \tag{2.12}$$

which is referred to as deformation gradient. As illustrated in Fig. 2.2 and computed in (2.11), the deformation gradient maps a line element connecting two material points found in an infinitesimal neighborhood in the referential to the current placement.

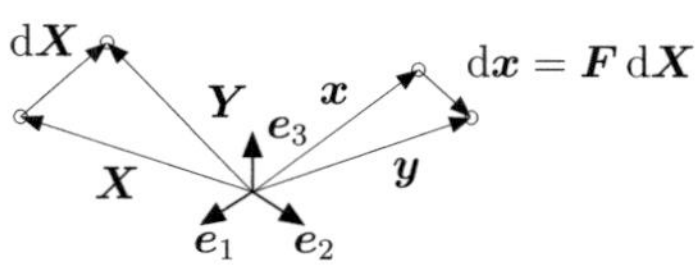

Figure 2.2: Transformation of line elements from the referential into the current placement

The displacement gradient $\boldsymbol{H}(\boldsymbol{X},t)$ is defined analogously to $\boldsymbol{F}(\boldsymbol{X},t)$ by

$$\boldsymbol{H}(\boldsymbol{X},t) = \mathrm{grad}_{\boldsymbol{X}}(\boldsymbol{u}(\boldsymbol{X},t)) = \boldsymbol{F}(\boldsymbol{X},t) - \boldsymbol{I}. \tag{2.13}$$

Then the change in the displacement of two points separated by the infinitesimal line element $\mathrm{d}\boldsymbol{X}$ is $\mathrm{d}\boldsymbol{u} = \boldsymbol{H}\,\mathrm{d}\boldsymbol{X}$. If the displacement gradient is sufficiently small, i.e. if

$$||\boldsymbol{H}(\boldsymbol{X},t)||_2 \ll 1 \tag{2.14}$$

holds, then a geometrically linear description of the deformation process is possible. The displacement gradient $\boldsymbol{H}$ can be decomposed into a symmetric part $\boldsymbol{\varepsilon}$ and a skew symmetric part $\boldsymbol{\omega}$

$$\boldsymbol{H}(\boldsymbol{X},t) = \boldsymbol{\varepsilon}(\boldsymbol{X},t) + \boldsymbol{\omega}(\boldsymbol{X},t), \tag{2.15}$$

with

$$\varepsilon(\boldsymbol{X},t) = \frac{1}{2}\left(\boldsymbol{H}(\boldsymbol{X},t) + \boldsymbol{H}(\boldsymbol{X},t)^{\mathsf{T}}\right) = \mathrm{sym}(\boldsymbol{H}(\boldsymbol{X},t)), \tag{2.16}$$

$$\omega(\boldsymbol{X},t) = \boldsymbol{H}(\boldsymbol{X},t) - \varepsilon(\boldsymbol{X},t) = \mathrm{skw}(\boldsymbol{H}(\boldsymbol{X},t)). \tag{2.17}$$

In the following, it is asserted that the condition (2.14) is satisfied. Then the skew symmetric part ω of the displacement gradient constitutes an infinitesimal rotation. The irrotational part ε is a measure for the pure distortion and dilatation omitting rigid body rotations. Note that rigid body translations do not enter into $\boldsymbol{H}$. The tensor ε is called the infinitesimal strain tensor. It can further be decomposed into a spherical part ε° and a deviatoric part ε' (Fig. 2.3):

$$\varepsilon^\circ = \frac{\mathrm{tr}(\varepsilon)}{3}\boldsymbol{I}, \tag{2.18}$$

$$\varepsilon' = \varepsilon - \varepsilon^\circ. \tag{2.19}$$

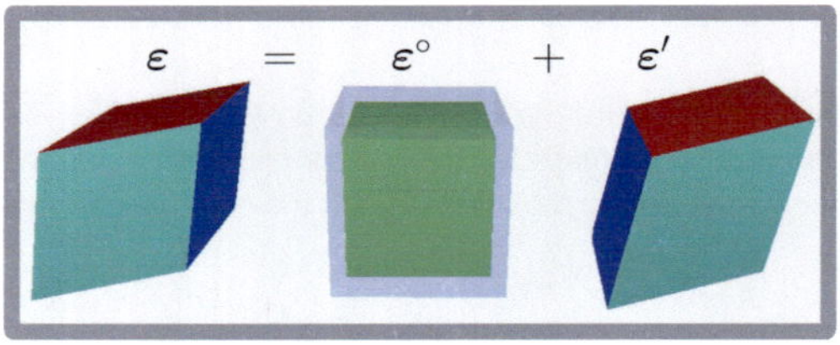

Figure 2.3: Decomposition of the total strain into spherical and deviatoric part

From the linearization of the relative change in volume

$$\mathrm{d}v = \det(\boldsymbol{F}(\boldsymbol{X},t))\,\mathrm{d}V \tag{2.20}$$

the relation

$$\mathrm{d}v = (1 + \mathrm{tr}(\varepsilon(\boldsymbol{X},t)))\,\mathrm{d}V \tag{2.21}$$

is obtained. Hence, ε° accounts for volumetric deformations, whereas ε' accounts for distortions which are isochoric.

2.2 Constitutive Modeling

2.2.1 Mechanical field variables and balance of momentum

In the following the mass density of the body is denoted ϱ. Under an external specific force field $\boldsymbol{b}$, e.g., a gravitational field, the force density $\varrho\boldsymbol{b}$ is exerted on the body. Inside of the body the Cauchy stress tensor $\boldsymbol{\sigma}$ is assumed. Couple stresses, body couples and multi-polar interactions as first introduced by the Cosserat brothers (Cosserat and Cosserat, 1909) are not accounted for, which is in contrast to other approaches, e.g., the one by Forest et al. (1997). More specifically the kinematic assertion is made that the deformation of material points is uniquely defined by their spatial translation, whereas in the Cosserat theory each material point is equipped with a director and its rotations is considers as an additional deformation mode. A continuum following these restrictions is denoted a Cauchy or a Boltzmann continuum. By virtue to Cauchy's lemma the traction vector $\boldsymbol{t}$ on a surface $\mathscr{S}$ with outward unit normal $\boldsymbol{n}$ is (Fig. 2.4)

$$\boldsymbol{t} = \boldsymbol{\sigma}\boldsymbol{n}. \tag{2.22}$$

Further, the stress tensor is symmetric

$$\boldsymbol{\sigma} = \boldsymbol{\sigma}^{\mathsf{T}} \tag{2.23}$$

as a consequence of balance of moment of momentum.

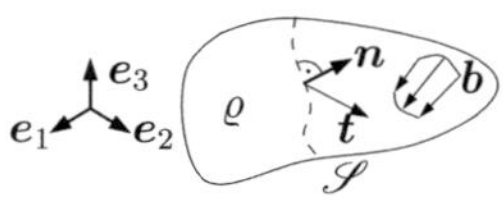

Figure 2.4: Illustration of the traction vector on a surface inside the body

The total mass of the body m and the linear momentum $\varrho\boldsymbol{v}$ are commonly assumed as conservatory variables. Taking into account the local form of the general balance equation of a conserved quantity $\psi(\boldsymbol{x}, t)$ with production p, supply s and flux $\boldsymbol{q}$

$$\frac{\partial\psi(\boldsymbol{x}, t)}{\partial t} + \operatorname{div}(\psi(\boldsymbol{x}, t)\boldsymbol{v}(\boldsymbol{x}, t)) = p(\boldsymbol{x}, t) + s(\boldsymbol{x}, t) + \operatorname{div}(\boldsymbol{q}(\boldsymbol{x}, t)) \tag{2.24}$$

and taking into account that the mass production p_{m}, the mass supply s_{m} and the inward mass flux $\boldsymbol{q}_{\mathrm{m}}$ are usually assumed to vanish in the field of solid mechanics, the balance of mass reads

$$\frac{\partial\varrho(\boldsymbol{x}, t)}{\partial t} + \operatorname{div}(\varrho(\boldsymbol{x}, t)\boldsymbol{v}(\boldsymbol{x}, t)) = 0. \tag{2.25}$$

Taking into consideration the representation (2.9) of the material time derivative, the balance of mass is rewritten as

$$\dot{\varrho}(\boldsymbol{x}(\boldsymbol{X}, t), t) + \varrho\operatorname{div}(\boldsymbol{v}) = 0. \tag{2.26}$$

The latter implies that for incompressible materials, i.e. for $\dot{\varrho} = 0$, the divergence of the velocity field has to vanish under the assumption of zero mass production, mass supply and non-convective mass flux. By taking $\boldsymbol{\sigma}$ as the flux of linear momentum and the force field $\varrho\boldsymbol{b}$ as the supply, the balance equation of the linear momentum

$$\boldsymbol{m}(\boldsymbol{x},t) = \boldsymbol{v}(\boldsymbol{x},t)\varrho(\boldsymbol{x},t) \tag{2.27}$$

can be reformulated into the well-known representation

$$\varrho(\boldsymbol{x},t)\boldsymbol{b}(\boldsymbol{x},t) + \mathrm{div}(\boldsymbol{\sigma}(\boldsymbol{x},t)) = \varrho(\boldsymbol{x},t)\ddot{\boldsymbol{x}}. \tag{2.28}$$

Assuming quasi-static processes, (2.28) further simplifies to

$$\varrho(\boldsymbol{x},t)\boldsymbol{b}(\boldsymbol{x},t) + \mathrm{div}(\boldsymbol{\sigma}(\boldsymbol{x},t)) = \boldsymbol{0}. \tag{2.29}$$

The partial differential equations (2.28) and (2.29) have to be augmented by suitable boundary conditions and, in the case of (2.28), by initial conditions. Usually, the following boundary conditions are distinguished

- *Kinematic boundary conditions*
  Prescribed displacements $\boldsymbol{u}_*$ on $\mathit{\Gamma}_\mathrm{u} \subseteq \mathit{\Gamma}$.

- *Static boundary conditions*
  Prescribed tractions $\boldsymbol{t}_*$ on $\mathit{\Gamma}_\mathrm{t} \subsetneq \mathit{\Gamma}$ ($\mathit{\Gamma}_\mathrm{t} = \mathit{\Gamma}$ is admissible for dynamic problems).

- *Mixed boundary conditions*
  Tractions $\boldsymbol{t}^*$ and displacements $\boldsymbol{u}^*$ are simultaneously defined. The prescribed fields have to satisfy the restriction $\boldsymbol{t}^* \cdot \boldsymbol{u}^* = 0$.

- *Combined boundary conditions*
  Prescribed tractions $\boldsymbol{t}_* = \boldsymbol{t}(\boldsymbol{u})$ on $\mathit{\Gamma}_\mathrm{c} \subseteq \mathit{\Gamma}$.

2.2.2 Thermodynamic field variables and energy balance

In order to describe thermodynamical processes of a body, additional variables incorporating thermodynamic information are required. First, the absolute temperature θ and the specific entropy η are considered in the following. Additionally, the specific internal energy e, the (outward) heat flux $\boldsymbol{q}$ and external heating h are accounted for. Here and thereafter it is allowed for a generalized vector of internal variables $\boldsymbol{\xi}$ containing additional information about the state of the material (see, e.g., Coleman and Gurtin, 1967). Examples for the latter are quantities describing the hardening state of a material (e.g., dislocation densities) or chemical concentrations. Then e does also depend on $\boldsymbol{\xi}$. The first law of thermodynamics states that the internal energy of a body is conserved

$$\begin{aligned}\frac{\partial\varrho(\boldsymbol{x},t)e(\boldsymbol{x},t)}{\partial t} &+ \mathrm{div}(\varrho(\boldsymbol{x},t)e(\boldsymbol{x},t)\boldsymbol{v}(\boldsymbol{x},t)) = \\ &h(\boldsymbol{x},t) + \boldsymbol{\sigma}(\boldsymbol{x},t)\cdot\dot{\boldsymbol{\varepsilon}}(\boldsymbol{x},t) - \mathrm{div}(\boldsymbol{q}(\boldsymbol{x},t)).\end{aligned} \tag{2.30}$$

Incorporating the balance of mass (2.30) is reorganized into the more convenient

$$\varrho \dot{e} = h + \boldsymbol{\sigma} \cdot \dot{\boldsymbol{\varepsilon}} - \operatorname{div}(\boldsymbol{q}). \tag{2.31}$$

In the following it is asserted that the internal energy is of the form

$$e = e(\boldsymbol{\varepsilon}, \dot{\boldsymbol{u}}, \theta, \eta, \boldsymbol{\xi}), \tag{2.32}$$

where the dependency on the velocity $\dot{\boldsymbol{u}}$ is negligible for quasi-static processes. In the following it is asserted that the supply and flux of the entropy η are given by

$$h_\eta = \frac{h}{\theta}, \qquad \boldsymbol{q}_\eta = \frac{\boldsymbol{q}}{\theta}. \tag{2.33}$$

The latter is motivated by the seminal works of Clausius (Clausius, 1850, 1865) who first introduced the second law of thermodynamics. A critical discussion of this constitutive assumption is presented in Müller (1985). In its original form it states that for any admissible thermodynamic cycle the inequality

$$\oint \frac{\mathrm{d}Q}{\theta} \leq 0 \tag{2.34}$$

has to hold, where Q is the total heat supplied to the observed body. Further, the special case of equality in (2.34) holds for reversible processes only. Then the rate of the total specific entropy production $\dot{\gamma}$ is

$$\varrho \dot{\gamma} = \varrho \dot{\eta} - \left(\frac{h}{\theta} - \operatorname{div}\left(\frac{\boldsymbol{q}}{\theta} \right) \right) = \varrho \dot{\eta} - \frac{\varrho \dot{e}}{\theta} + \frac{1}{\theta} \boldsymbol{\sigma} \cdot \dot{\boldsymbol{\varepsilon}} - \frac{\operatorname{grad}(\theta) \cdot \boldsymbol{q}}{\theta^2} \tag{2.35}$$

The Clausius-Duhem inequality derived from (2.34) asserts that the rate of entropy production cannot be negative

$$\dot{\gamma} \geq 0. \tag{2.36}$$

By virtue of the Legendre Fenchel transformation the internal energy e can be expressed in terms of the Helmholtz free energy ψ, the entropy η and the absolute temperature θ

$$e(\boldsymbol{x}, t) = \psi(\boldsymbol{x}, t) + \theta(\boldsymbol{x}, t)\eta(\boldsymbol{x}, t). \tag{2.37}$$

By making use of (2.37) in (2.35) and multiplying the result by the non-negative absolute temperature θ, the energy balance (2.31) can be used to rewrite (2.36)

$$\varrho \dot{\gamma} = -\varrho \dot{\psi} - \varrho \eta \dot{\theta} + \boldsymbol{\sigma} \cdot \dot{\boldsymbol{\varepsilon}} - \frac{\operatorname{grad}(\theta) \cdot \boldsymbol{q}}{\theta} \geq 0. \tag{2.38}$$

The following implications can be drawn:

- The rate of internal entropy $\dot{\gamma}$ is non-negative whenever $\dot{\boldsymbol{\varepsilon}} = \mathbf{0}, \dot{e} = 0, \mathrm{grad}(\theta) = \mathbf{0}$.
- The internal energy is monotonic decreasing ($\dot{e} \leq 0$) for $\dot{\boldsymbol{\varepsilon}} = \mathbf{0}, \dot{\gamma} = 0, \mathrm{grad}(\theta) = \mathbf{0}$.
- The free energy is monotonic decreasing ($\dot{\psi} \leq 0$) for $\dot{\boldsymbol{\varepsilon}} = \mathbf{0}, \dot{\theta} = 0, \mathrm{grad}(\theta) = \mathbf{0}$.

Incorporating $\psi \equiv \psi(\boldsymbol{\varepsilon}, \theta, \boldsymbol{\xi})$ one can write

$$\varrho\dot{\gamma} = \left(\boldsymbol{\sigma} - \varrho\frac{\partial\psi}{\partial\boldsymbol{\varepsilon}}\right) \cdot \dot{\boldsymbol{\varepsilon}} - \left(\varrho\frac{\partial\psi}{\partial\theta} + \varrho\eta\right)\dot{\theta} - \varrho\frac{\partial\psi}{\partial\boldsymbol{\xi}} \cdot \dot{\boldsymbol{\xi}} - \frac{\mathrm{grad}(\theta) \cdot \boldsymbol{q}}{\theta} \geq 0. \tag{2.39}$$

For reversible processes, i.e., for elastic, isothermal cycles, the entropy production is zero and, hence,

$$\boldsymbol{\sigma} = \varrho\frac{\partial\psi}{\partial\boldsymbol{\varepsilon}} \tag{2.40}$$

has to hold. Considering a homogeneous body undergoing arbitrary temperature rates at constant deformation and internal state, it is found that the entropy is related to the free energy via

$$\eta = -\frac{\partial\psi}{\partial\theta}. \tag{2.41}$$

It is common to refer to $\boldsymbol{\sigma}$ and η as the thermodynamic conjugate variables of $\boldsymbol{\varepsilon}$ and θ, respectively. Analogously, the generalized vector of thermodynamic forces

$$\boldsymbol{\mathcal{F}} = -\varrho\frac{\partial\psi}{\partial\boldsymbol{\xi}} \tag{2.42}$$

is introduced. Accounting for (2.40) and (2.41), the Clausius-Duhem inequality (2.39) simplifies to the compact dissipation inequality

$$\mathcal{D} = \boldsymbol{\mathcal{F}} \cdot \dot{\boldsymbol{\xi}} - \frac{\boldsymbol{g} \cdot \boldsymbol{q}}{\theta} \geq 0. \tag{2.43}$$

Subdividing the dissipation into a mechanical contribution $\mathcal{D}_{\mathrm{m}}$ and a thermal part $\mathcal{D}_\theta$, and substituting $\boldsymbol{g} = \mathrm{grad}(\theta)$ the reduced Clausius-Duhem inequality becomes

$$\boldsymbol{\mathcal{F}} \cdot \dot{\boldsymbol{\xi}} - \frac{\boldsymbol{g} \cdot \boldsymbol{q}}{\theta} = \mathcal{D}_{\mathrm{m}} + \mathcal{D}_\theta \geq 0. \tag{2.44}$$

Often it is assumed that the mechanical and thermal dissipation are independent and, thus, two individual inequalities are obtained

$$\mathcal{D}_{\mathrm{m}} = \boldsymbol{\mathcal{F}} \cdot \dot{\boldsymbol{\xi}} \geq 0, \qquad \mathcal{D}_\theta = -\frac{\boldsymbol{q} \cdot \boldsymbol{g}}{\theta} \geq 0. \tag{2.45}$$

The latter imposes restrictions on the evolution of the internal variables and is frequently employed in order to verify the thermodynamic consistency of material models.

Notably, the non-negativity of the thermal dissipation implies

$$-\boldsymbol{g} \cdot \boldsymbol{q} \geq 0. \tag{2.46}$$

Assuming $\boldsymbol{q} \equiv \boldsymbol{q}(-\boldsymbol{g})$ and setting $\boldsymbol{q}(\boldsymbol{0}) = \boldsymbol{0}$, one can express $\boldsymbol{q}$ in terms of the Taylor expansion

$$\boldsymbol{q} = \underbrace{\boldsymbol{q}(\boldsymbol{0})}_{=\boldsymbol{0}} - \boldsymbol{\kappa}\boldsymbol{g} + \frac{1}{2}\boldsymbol{\kappa}_3[\boldsymbol{g} \otimes \boldsymbol{g}] + \dots . \tag{2.47}$$

Due to the inequality (2.46) the tensor $\boldsymbol{\kappa}$ has to be symmetric and positive, if non-negative thermal dissipation is assumed. When neglecting contributions of second and higher order in $\boldsymbol{g}$, and assuming $\boldsymbol{\kappa}$ to be constant the classical Fourier law of heat conductivity is obtained

$$\boldsymbol{q} = -\boldsymbol{\kappa}\boldsymbol{g}. \tag{2.48}$$

Using (2.37), (2.40), (2.41) and (2.42), the energy balance (2.31) can be reformulated as

$$-\varrho\theta \frac{\partial^2 \psi}{\partial \theta^2}\dot{\theta} = \theta \frac{\partial \boldsymbol{\sigma}}{\partial \theta} \cdot \dot{\boldsymbol{\varepsilon}} + h + \left(\boldsymbol{\mathcal{F}} - \theta \frac{\partial \boldsymbol{\mathcal{F}}}{\partial \theta} \right) \cdot \dot{\boldsymbol{\xi}} - \operatorname{div}(\boldsymbol{q}). \tag{2.49}$$

The representation (2.49) is used in the following.

2.2.3 Generalized Standard Materials

The postulation of material models is a complex procedure and only few very general rules are obtained from the previous investigation. In particular, the structure of the constitutive equations is mostly unclear. Originating from the French mechanics community (Halphen and Nguyen, 1975; Germain et al., 1983) the notion of generalized standard materials (GSM) was introduced. Its aim is to provide a clear guideline and mathematical tools for the construction of constitutive equations for material models. A generalized standard material is described by two potentials, (i) the Helmholtz free energy ψ (2.37) and (ii) the dissipation potential ϕ. The latter is a function of the rate of internal variables $\dot{\boldsymbol{\xi}}$. For such materials the thermodynamic conjugate forces $\boldsymbol{\mathcal{F}}$ are

$$\boldsymbol{\mathcal{F}} = \frac{\partial \phi(\dot{\boldsymbol{\xi}})}{\partial \dot{\boldsymbol{\xi}}}. \tag{2.50}$$

An alternative expression is obtained from the dual dissipation potential

$$\phi^*(\boldsymbol{\mathcal{F}}) = \sup_{\dot{\boldsymbol{\xi}}} \left\{ \boldsymbol{\mathcal{F}} \cdot \dot{\boldsymbol{\xi}} - \phi(\dot{\boldsymbol{\xi}}) \right\}, \qquad \dot{\boldsymbol{\xi}} = \frac{\partial \phi^*}{\partial \boldsymbol{\mathcal{F}}}, \tag{2.51}$$

which is obtained by means of the Legendre transformation. It can be constructed if ϕ is convex. Generalized standard materials can easily be employed in a variational formulation of the constitutive equations (e.g., Hackl, 1996; Ortiz and Repetto, 1999). A thermodynamical coupled variational representation was introduced by Yang et al. (2006) and recently extended by Stainier and Ortiz (2010). While the presented formulas suggest to be restricted to rate-dependent materials, e.g., to the visco-plastic case, rate-independent behavior is covered within the framework by setting

$$\phi^*(\boldsymbol{\mathcal{F}}) = \begin{cases} 0 & \varphi(\boldsymbol{\mathcal{F}}) \leq 0, \\ \infty & \text{else}, \end{cases} \tag{2.52}$$

with $\varphi(\boldsymbol{\mathcal{F}})$ being a level-set function denoting the yield surface in the case of plasticity or the damage surface for the investigation of material failure. The evolution of internal variables is then determined by introducing Lagrangian multipliers enforcing the constraint $\varphi(\boldsymbol{\mathcal{F}}) \leq 0$.

In the case of a rate-independent von Mises plasticity model (von Mises, 1928) with isotropic hardening, the set of internal variables is $\boldsymbol{\xi} = \{\boldsymbol{\varepsilon}^{\mathrm{p}}, q\}$ consists of the infinitesimal plastic strain tensor $\boldsymbol{\varepsilon}^{\mathrm{p}}$ and the scalar hardening variable q. The generalized vector of conjugate forces $\boldsymbol{\mathcal{F}}$ then becomes

$$\boldsymbol{\mathcal{F}} = (\boldsymbol{\sigma}, -r)^{\mathsf{T}}. \tag{2.53}$$

The admissible range of the stress tensor is defined by the condition

$$\varphi(\boldsymbol{\sigma}, r) = \|\boldsymbol{\sigma}'\|_2 - \sqrt{\frac{2}{3}}(\sigma_{\mathrm{F0}} + r) \leq 0, \tag{2.54}$$

which defines a dual dissipation potential ϕ^* of the form (2.52). The condition (2.54) is a function depending only on the vector of conjugate forces and it determines the rate of internal variables

$$\dot{\boldsymbol{\varepsilon}}^{\mathrm{p}} = \dot{\lambda}\frac{\boldsymbol{\sigma}'}{\|\boldsymbol{\sigma}'\|_2}, \qquad \dot{q} = \dot{\lambda}\sqrt{\frac{2}{3}}, \tag{2.55}$$

with the non-negative Lagrangian multiplier $\dot{\lambda} \geq 0$.

Chapter 3

Real microstructures and model representations

3.1 Motivation

In this work methods for the numerical prediction of the overall thermo-mechanical properties of materials with microstructure are investigated. Such microheterogeneous materials show a constitutive behavior that is strongly depending on a variety of interacting parameters. In particular, the macroscopic thermo-mechanical response depends on the topology and the micromorphology of the material on one or several smaller scales. An

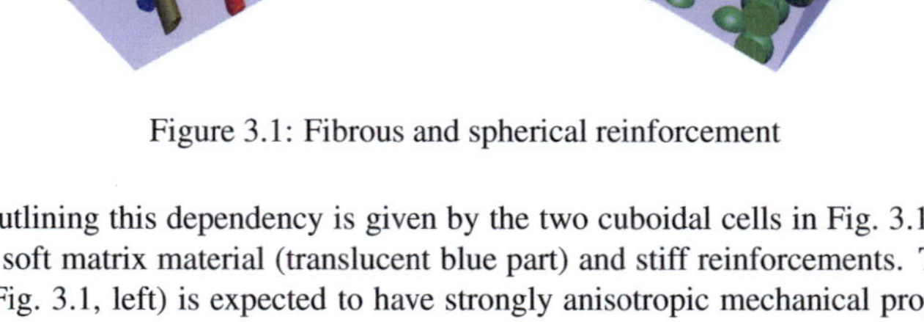

Figure 3.1: Fibrous and spherical reinforcement

example outlining this dependency is given by the two cuboidal cells in Fig. 3.1 which are made of a soft matrix material (translucent blue part) and stiff reinforcements. The fibrous material (Fig. 3.1, left) is expected to have strongly anisotropic mechanical properties due to the near uniform fiber orientation, whereas the material containing spherical particles (Fig. 3.1, left) is expected to exhibit an almost isotropic mechanical behavior.

The presented examples emphasize the importance that comes to the consideration of the actual microscopic geometry. A quantitative description of the latter is provided in section 3.2 based on a selection of measures for the statistical characterization. Emphasis is placed on the characterization of crystalline aggregates, particulate and porous materials. Additionally, the general concept of the n-point correlation function is briefly introduced. Comprehensive overviews on statistical descriptions of microstructures are, e.g., given in the books of Ohser and Mücklich (2000); Torquato (2002).

Based on experimental data and the presented statistical descriptions, microstructures can be quantitatively characterized and compared to each other, e.g., in order to investigate the effect of morphological changes on the effective properties of the material. However, the necessary experimental observations performed on the relevant small scales (usually in the micrometer regime) are complicated by several factors. Experimental techniques commonly applied include serial sectioning in combination with electron back-scatter diffraction (EBSD) for polycrystals, X-ray and neutron diffraction methods and the popular (micro-)computer-tomography (e.g., Dillard et al., 2005; Madi et al., 2007). First, the preparation and acquisition of the specimens is often a complicated procedure. Therefore, experimental costs are often high and, moreover, the time needed to prepare the specimens can be substantial. This holds particularly true, if many different materials in small quantiles are envisaged. Second, the gathered experimental data is subjected to a large number of perturbations. Data post-processing is, hence, an often indispensable step. Common operations at this stage involve noise reduction, smoothing, filtering, binarization and segmentation (see, e.g., Gonzalez and Woods, 2002; Jähne, 2005). Third, the obtained information can usually not be used in numerical simulations in a straight-forward way. This is due to the pronounced degree of non-uniformity of the materials accompanied by the aforementioned binarization issues. These factors hinder the generation of spatial discretizations which are a prerequisite for many computational methods, namely for the popular finite element method (see, e.g., Bathe, 2002; Zienkiewicz et al., 2006).

Following the ever-increasing demand in fast and accurate methods for the prediction of material properties, computationally efficient geometric representations of different classes of materials are required. With respect to the last of the aforementioned points, an exact description of the microstructure based on purely experimental evidence is unlikely to be successful, although modern image processing and mesh generation software (e.g., AVIZO*, or ScanIP/ScanFE†) offer increasing capabilities. An alternative for some applications can be found in voxel based schemes using either the finite element method in combination with multi-phase elements (e.g., Zhodi and Wriggers, 2005) or the fast fourier transform (Moulinec and Suquet, 1998; Lebensohn et al., 2008). However, these techniques have

*Visualization Sciences Group, `http://www.vsg3d.com`

†Simpleware, `http://www.simpleware.com`

other disadvantages such as large data sets, inexact geometric representation and restricted boundary conditions. Moreover, no reliable data close to the interface between different materials can be obtained from these methods, which partially precludes the methods from application to damage initiation and nucleation. These techniques are not pursued in the following.

In order to circumvent the aforementioned problems and to facilitate the mesh generation, the concept of model microstructures has shown to be an efficient tool for the prediction of the behavior of microheterogeneous materials (see, e.g., Kumar and Kurtz, 1994; Decker and Jeulin, 2000; Barbe et al., 2001; Kanit et al., 2003; Osipov et al., 2008; Fritzen et al., 2009; Fritzen and Böhlke, 2011c). The key idea is to replace the real microstructure by an artificial one with similar statistical properties. Based on the statistical description addressed in section 3.2, the difference of the approximation and real specimens can be evaluated. The artificial structures are often generated using geometric primitives such as, e.g., spheres, cylinders or polyhedra, in order to approximate the real microstructure. The geometric primitives can be assembled, e.g., by applying affine transformations (translations/rotations) to the prototypes or by the combination of different geometric bodies using boolean operators (AND/OR/NOT/XOR). Usually, the discrete geometry is parametrized using a set of variables $\mathcal{G} \subset \mathbb{R}^n$ (n: number of parameters). In order to investigate the influence of scattering of the microstructure, the parameter vector $\mathcal{G}$ can be expressed as a function of random variables $Z = (Z_1, \ldots, Z_n)$ with prescribed distribution. Fig. 3.2 illustrates the general procedure of constructing an artificial microstructure based on the proposed scheme. Besides the improved spatial discretization possibilities of the model microstructures, the artificial materials can, further, be used in design studies in order to systematically evaluate the effect of changes in the local topology and/or morphology of materials onto their effective behavior. The latter is an interesting feature with regard to the arising scientific field of so called *tailored materials* (e.g., Kumar and McDowell, 2009). An example for such studies is the explicit investigation of the effect of the regularity of the particle shape on the effective thermo-elastic properties of metal ceramic composites (MMC) presented by Fritzen and Böhlke (2011c). Another approach in this direction was pursued by Fritzen and Böhlke (2011b) to find the influence of anisotropic particle morphology of MMCs on the physically nonlinear effective response of composite materials.

Different classes of model microstructures are presented in the remainder of the chapter for the consideration of crystalline aggregates (section 3.3.1), materials containing polyhedral particles (section 3.3.2) and spherical pores (section 3.4). Anisotropic modifications of some of the proposed microstructures are investigated. Aspects of three-dimensional high quality spatial discretizations using finite elements are presented, with particular focus on the periodicity of the created meshes.

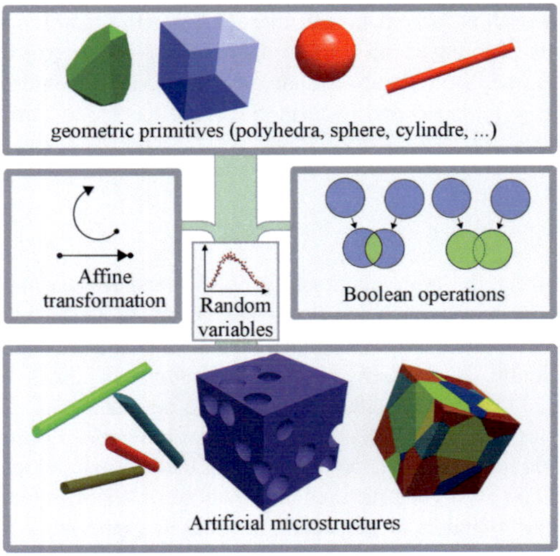

Figure 3.2: Schematic representation of a general algorithm for the generation of artificial microstructures involving random variables Z

3.2 Statistical properties of microstructures

In the following, statistical homogeneity is asserted. Mathematically this means that when examining a microstructure, the local properties are distributed according to a spatially constant probability density. In particular, the same statistical distribution of the properties is found at any position. A counterexample to this are graded microstructure, for which the distribution function varies with respect to the spatial position.

For statistically homogeneous materials all individual regions contained in a single random realization (of infinite size) of the material are statistically similar. For such materials it is sensible to propose an ergodicity hypothesis. Under this hypothesis ensemble averaging, i.e. taking the expected value of all possible realization of the microstructure at a given point, is equivalent to volume averaging in the infinite volume limit (see also, Torquato, 2002). In praxis this states that the average over arbitrarily chosen points in a material is equivalent to taking the mean of the same number of random specimens at the same position, if an infinite number of points is considered.

The characterization of microstructures can be pursued using a variety of techniques. As mentioned earlier, profound overviews on the topic are given by, e.g., Ohser and Mücklich (2000); Torquato (2002). The following mathematical descriptions refer to the discrete indicator function $\chi_i(\boldsymbol{X})$ of the i-th component*. The indicator function equals one, if the point $\boldsymbol{X}$ is found in the i-th component. Otherwise, it is zero.

3.2.1 Microstructural n-point statistics

The n-point auto-correlation of the component i is a mapping from $\mathbb{R}^3 \times \cdots \times \mathbb{R}^3$ (n-times) on the real interval $[0, 1]$ which is defined via

$$f_i^{(n)}(\boldsymbol{r}_1, \ldots, \boldsymbol{r}_n) = \frac{1}{|\Omega|} \int_{\Omega} \left(\prod_{j=1,\ldots,n} \chi_i(\boldsymbol{X} + \boldsymbol{r}_j) \right) \mathrm{d}V. \tag{3.1}$$

Due to the asserted statistical homogeneity the n-point auto-correlation is independent of the position vector $\boldsymbol{r}_1$, i.e., and fixing $\boldsymbol{r}_1 = \boldsymbol{0}$ does not alter the result. An interpretation of the n-point function is the probability of finding for a given set of constant vectors $\boldsymbol{r}_j$ $(j = 2, \ldots, n)$ the same (i-th) material at a random point $\boldsymbol{X}$ and at all points $\boldsymbol{X} + \boldsymbol{r}_j$. An equivalent and more illustrative description can be obtained by introduction of the subdomain $\Omega_i \subseteq \Omega$ occupied by the i-th component. For simplicity the domain Ω_i translated by the vector $\boldsymbol{r}_j$ is denoted by $\Omega_i \oplus \boldsymbol{r}_j$. Then the n-point function can be expressed by $(\boldsymbol{r}_1 = \boldsymbol{0})$

$$f_i^{(n)}(\boldsymbol{r}_1, \ldots, \boldsymbol{r}_n) = \frac{1}{|\Omega|} \left| \bigcap_{j=1,\ldots,n} (\Omega_i \oplus \boldsymbol{r}_j) \right| . \tag{3.2}$$

The formulation (3.2) allows for the following implications:

- The 1-point function determines the concentration c_i of the component within the reference volume:

$$f_i^{(1)} = \frac{|\Omega_i|}{|\Omega|} = c_i. \tag{3.3}$$

- For $m > n \geq 1$ the identities

$$f_i^{(m)}(\boldsymbol{r}_1 = \boldsymbol{0}, \ldots, \boldsymbol{r}_n, \boldsymbol{0}, \ldots, \boldsymbol{0}) = f_i^{(n)}(\boldsymbol{r}_1, \ldots, \boldsymbol{r}_n), \tag{3.4}$$

$$\Rightarrow f_i^{(m)}(\boldsymbol{0}, \ldots, \boldsymbol{0}) = f_i^{(1)}(\boldsymbol{0}) = c_i. \tag{3.5}$$

  hold. Moreover, the inequality

$$f_i^{(m)}(\boldsymbol{r}_1, \ldots, \boldsymbol{r}_n, \boldsymbol{r}_{n+1}, \ldots, \boldsymbol{r}_m) \leq f_i^{(n)}(\boldsymbol{r}_1, \ldots, \boldsymbol{r}_n) \tag{3.6}$$

  is satisfied for arbitrary vectors $\boldsymbol{r}_{n+1}, \ldots, \boldsymbol{r}_m \in \mathbb{R}^3$.

*In the case of polycrystals each crystallographic orientation is treated as a separate *component*.

Many microstructures can be considered to be statistically near isotropic. For such microstructures the n-point correlation function is not depending on the direction of $\boldsymbol{r}_i$ but only on the distance $r_i = \|\boldsymbol{r}_i\|_2$. For such microstructures the computation of the n-point probability function from data sets can be improved considerably. However, the aspects of statistically anisotropic microstructures are ultimately of interest for the consideration of real materials and no statistical isotropy is asserted in the following.

In computational practice, the n-point correlation function is often evaluated on a binarized voxel representation of the microstructure at finite resolution. Most notably, the computational time for the evaluation is usually prohibitively high. More precisely it is of order $\mathcal{O}(N^{3n})$, where N is the resolution of the voxel image in one direction. Therefore, the n-point function is often evaluated in a small neighborhood, i.e., for bounded vectors $\boldsymbol{r}_i$ $(i = 2, 3, \dots)$, to reduce the computational cost. Third and higher-order statistical information is rarely computed due to the computational inefficiency and the difficult interpretation of the results. More specifically, the number of discrete values is squared when moving from the two-point to the three-point correlation function. Even for moderate spatial resolutions such as $N = 100$ the memory requirements largely exceed current standards. Due to the resemblance of the two point function to the convolution integral, the computation can be carried out with aid of the fast fourier transformation (FFT), if periodic microstructures are investigated. In practice FFT based methods are often used for non-periodic structures due to the superior performance and the boundary values are neglected.

3.2.2 Topology dependent statistical descriptions

The previously introduced n-point correlation functions allow for a general description of microstructures since no additional information on the material is required. However, the interpretation and the computation of the n-point function is complicated due to its generality. When considering a certain type of microstructure it is often possible to find a specific description that accounts for the micro-topology of the material, e.g., for fibers or particles. These descriptions have the favorable property of facilitated experimental or numerical evaluation, if chosen appropriately. Moreover, the interpretation of the measures is often simplified. A brief selection of topology dependent statistical descriptions are presented in the following.

3.2.2.1 Materials with convex granular structure

Materials with convex granular topology form an important class of microstructure. Examples are polycrystalline aggregates of metallic and ceramic materials or particles used in classical composites. In the following the terms grain, crystal and particle denote the individual microscopic entities. In order to characterize the morphology of individual grains, the following statistical measures are introduced:

- The volume fraction $f_i^{(1)} = c_i = V_i/V$, with $V_i = \Omega_i$ the volume occupied by the grain and V the total volume of the aggregate.

- The maximum grain length, minimum grain length and the mean grain diameter

$$l_i^{\max} = \max_{\boldsymbol{p},\boldsymbol{q} \in \Omega_i} \|\boldsymbol{p} - \boldsymbol{q}\|_2, \tag{3.7}$$

$$l_i^{\min} = \min_{\|\boldsymbol{n}\|_2 = 1} \max_{\boldsymbol{p},\boldsymbol{q} \in \Omega_i} |(\boldsymbol{p} - \boldsymbol{q}) \cdot \boldsymbol{n}|, \tag{3.8}$$

$$\bar{d}_i = 2\left(\frac{3V_i}{4\pi}\right)^{1/3}. \tag{3.9}$$

For the minimum and maximum grain length (also: minimum and maximum Feret), the directors $\boldsymbol{n}^{\max}, \boldsymbol{n}^{\min}$ in which the respective extremal values are found can also be determined.

The consideration of special grain topologies such as polyhedra gives rise to additional statistical parameters, e.g., the number of corners or edges, the mean edge length or the mean facet angle. For example, Kumar and Kurtz (1995) performed a significant number of these specific analysis.

The morphological grain anisotropy is an important influence factor on the thermo-mechanical properties of microheterogeneous materials. Its presence is often due to specific manufacturing processes such as the rolling process for sheet metal. It results in a grain elongation in one direction, a grain compaction in the thickness direction and approximately constant grain dimension in the transverse direction. In order to characterize such mean grain shape, one possible approach is to determine for a fixed triad of orthonormal director vectors $\boldsymbol{n}^{(\alpha)}$ $(\alpha = 1, 2, 3)$ the grain dimension

$$l_i(\boldsymbol{n}^{(\alpha)}) = \max_{\boldsymbol{p},\boldsymbol{q} \in \Omega_i} |(\boldsymbol{p} - \boldsymbol{q}) \cdot \boldsymbol{n}^{(\alpha)}|. \tag{3.10}$$

This is equivalent to the determination of the cuboidal envelope of the grain with respect to the given orthonormal basis $\{\boldsymbol{n}^{(\alpha)}\}$. If the directors are unknown, i.e. if there is no information on a possible specimen symmetry, the directors can enter in the optimization problem as additional unknowns. Other measures characterizing the anisotropy of the grain shape are the ratios of pairs of the maximum or minimum grain length and the mean grain diameter. A perfectly isotropic grain, i.e. a sphere, satisfies $l^{\min} = l^{\max} = \bar{d}$. As an example, the minimum and maximum length of a cuboid of non-dimensional length λ and width and height $1/\sqrt{\lambda}$ are considered in Fig. 3.3 for $\lambda \in [0.25, 4]$.

3.2.2.2 Porous materials

k-nearest neighbor distance. In the following a material containing N spherical non-intersecting pores of constant radius r is considered. A quantification of the pore-pore interactions in these materials is a difficult undertaking. In order to investigate the inhomogeneity of the pore topology, the k-nearest neighbor distance is analyzed (Holmes and Adams, 2002).

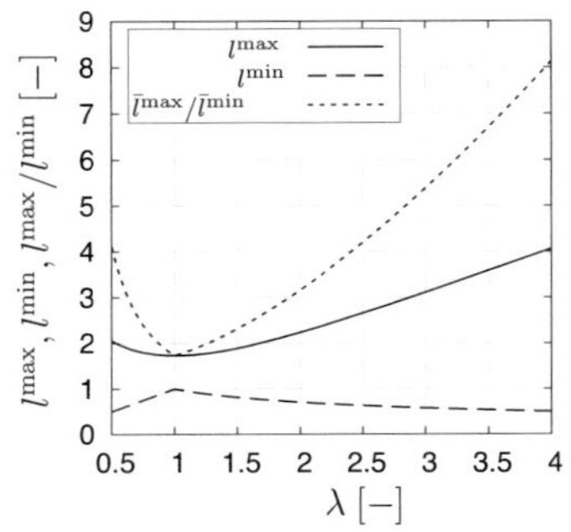

Figure 3.3: Example of the statistical measures $l^{max}, l^{min}, l^{max}/l^{min}$ for a cube of unit volume

It is defined using the center points of the pores $\boldsymbol{p}_\alpha \in \Omega$ $(1 \leq \alpha \leq N)$ via

$$d_k^\alpha = \arg\min_{d \in \mathbb{R}_+} \left\{ \text{card}\left(\{\beta \in \mathbb{N} \setminus \{\alpha\} : \|\boldsymbol{p}_\alpha - \boldsymbol{p}_\beta\|_2 \leq d + 2r\} \right) = k \right\}, \tag{3.11}$$

$$\bar{d}_k = \frac{1}{N} \sum_{\alpha=1}^{N} d_k^\alpha. \tag{3.12}$$

Comparison of $\bar{d}_k$ with the minimum $\check{d}_k$ and maximum $\hat{d}_k$ value over all particles can indicate the spread of the pore distance. For example, the minimum value of d_1^α for $\alpha \in \{1, \ldots, N\}$ indicates the minimum distance between neighboring pores, the maximum value determines the furthest distance between two pores and $\bar{d}_1$ is the average nearest neighbor distance.

Pore clustering. In order to judge on the clustering of pores, the k-nearest neighbor distance is of partial use. If more detailed information on particle or pore clusters are investigated, other statistical measures are required. A method used, e.g., by Bilger et al. (2005) is to increase the pore size and to find interconnected regions in the resulting virtual microstructure. For spherical topologies this is possible via an increment $\delta r > 0$ of the radius of all pores. The method is exemplified in Fig. 3.4 for the two-dimensional case.
Notice that for periodic topologies in three spatial dimensions the identification and visualization of pore clusters is more involved. Examples for the three-dimensional case can be found in section 3.4.

Two-point correlation function. As mentioned earlier, the computation of the two point function can be a challenging procedure. In the case of non-overlapping spheres it is, however, possible to evaluate the two-point correlation function by explicit computation of the volume of the intersection of translated spheres in order to compute the auto-correlation function precisely at selected points.

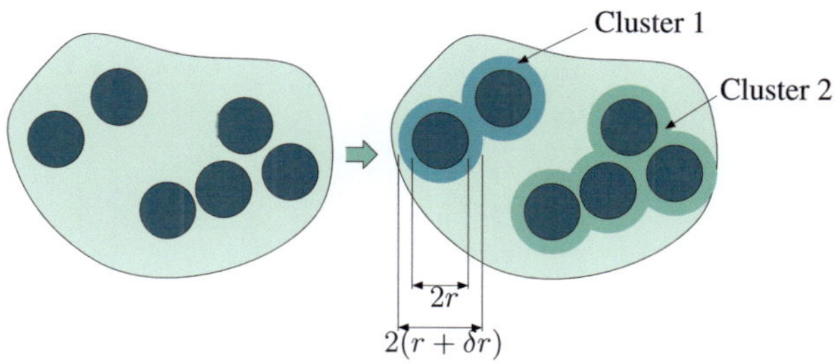

Figure 3.4: Illustration of the pore cluster identification algorithm (two-dimensional case)

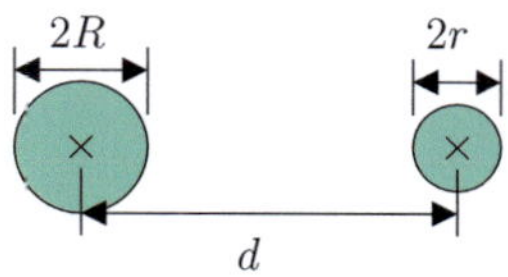

Figure 3.5: Geometric entities describing two spheres relative to each other

Therefore, the volume of the intersection of two spheres with radii r, R ($R \geq r$) and the center point distance d are analyzed (Fig. 3.5) in order to compute (3.2). The volume v of the intersection of these spheres can be computed by consideration of four distinct cases:

- The spheres do not intersect ($d \geq R + r$): $v = 0$ (see Fig. 3.6, case 1).
- The small sphere is found within the larger one ($d + r \leq R$): $v = 4\pi r^3/3$ (see Fig. 3.6, case 2).
- The spheres intersect and the intersection is found between the center points ($\sqrt{R^2 - r^2} \leq d < R + r$) (see Fig. 3.6, case 3). Then the distance of the center of the intersecting circle is found at the distance

$$a = \frac{R^2 - r^2 + d^2}{2d} \tag{3.13}$$

from the center of the larger sphere. Setting $b = d - a$ the sought-after volume of the intersection is

$$\begin{aligned} v &= \pi \int_a^R R^2 - \xi^2 \, \mathrm{d}\xi + \pi \int_b^r r^2 - \xi^2 \, \mathrm{d}\xi \\ &= \frac{\pi}{3} \left((R-a)^2 (2R + a) + (r - b)^2 (2r + b) \right) \end{aligned} \tag{3.14}$$

- The center of the intersection lies outside of the connection between the center points ($R - r < d < \sqrt{R^2 - r^2}$). The distance a (see Fig. 3.6, case 4) is

$$a = \frac{R^2 - r^2 - d^2}{2d}, \qquad b = d + a. \tag{3.15}$$

The volume of the intersection computes to

$$\begin{aligned} v &= \pi \int_{-r}^{a} r^2 - \xi^2 \,\mathrm{d}\xi + \pi \int_{b}^{R} R^2 - \xi^2 \,\mathrm{d}\xi \\ &= \frac{\pi}{3} \left((r^2 - a^2)(2r - a) + (R - b)^2 (2R + b) \right). \end{aligned} \tag{3.16}$$

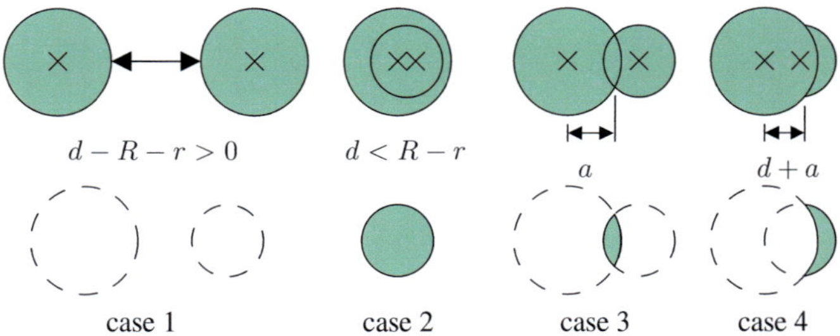

Figure 3.6: Computation of the intersection of two spheres: cases 1-4

Based on this algorithm it is possible to compute the two-point function *exactly* at selected points, e.g., on a regular grid. The obtained data can be post-processed in various programs. The presented algorithm is implemented in a C++ program in order to investigate the two-point correlation of pores in section 3.4. A multi-threaded implementation allows for parallel execution on shared memory multiprocessor systems. Near optimal scaling with respect to the number of processors was achieved. In particular, the use of N_{cpu} processors reduces the computation time to $1/N_{\text{cpu}}$.

3.3 Artificial microstructures based on random Voronoi tessellations

3.3.1 Representation of crystal aggregates

Many metallic or ceramic materials show a granular microstructure. The different grains are usually identified by their respective crystallographic orientation, which is often discontinuous at the grain boundary. Moreover, the orientation within individual grains after annealing can be considered almost constant. It is of interest to find a simple but suitable approximation of this granular topology in order to model metallic materials for the simulation of the mechanical behavior of the aggregate. The illustrations in this sections follow the line of Fritzen et al. (2009).

As pointed out by Kumar and Kurtz (1994, 1995) and others, the Voronoi tessellation is a powerful tool for the description of the almost polyhedral grains in a crystalline aggregate. The Voronoi tessellation is a nearest neighbor diagram determined from a set of N_g points commonly referred to as Voronoi generators. A point $\boldsymbol{x} \in \Omega$ of the microstructure belongs to the i-th grain, where i is defined by

$$i \in \mathcal{I} = \underset{1 \leq j \leq N_g}{\operatorname{argmin}} \left(\| \boldsymbol{x} - \boldsymbol{p}_j \|_2 \right) . \tag{3.17}$$

This definition is unique for interior points, i.e., $|\mathcal{I}| = 1$. For a non-degenerated Voronoi tessellation points found on the faces connecting neighboring grains satisfy $|\mathcal{I}| = 2$. Further, $|\mathcal{I}| = 3$ holds on edges and at corners $|\mathcal{I}| = 4$ is found. See Aurenhammer (1991) for a comprehensive review on Voronoi tessellations and their application.

A convenient property of the thereby generated diagram is the convexity of the cells and the possible incorporation of a periodicity constraint. The latter can be realized by copying the Voronoi generator point seed $\mathcal{S}$ around the (for simplicity cuboidal) unit cell as is illustrated in two space dimensions in Fig. 3.7. In three space dimensions this results in 26 translated copies of the original N_g generator points. The number of points for which the actual Voronoi tessellation has to be constructed is hence 27 times the number of cells which are to be generated when periodicity constraints are imposed.

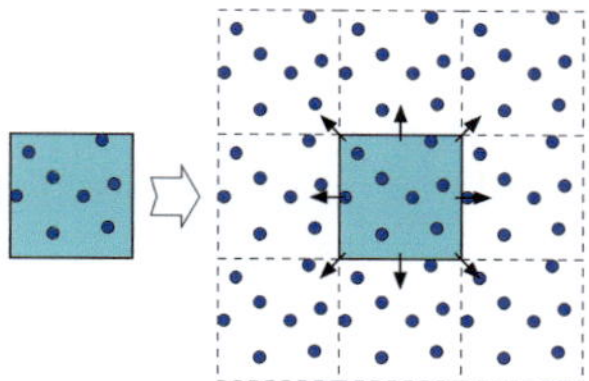

Figure 3.7: Construction of a periodic set of Voronoi vertices for implication of a periodic microstructure (cf. Fritzen et al., 2009)

The representation of a polycrystal by a Voronoi diagram is motivated by the following assumptions (see also Kumar and Kurtz, 1994, 1995):

- Grain growth starts at all points $\boldsymbol{p}_i$ in a finite set of nuclei $\mathcal{S} \subset \mathbb{R}^3$ at the same time T_0. The nuclei are fixed at their spatial position during the growth process, i.e. they do not move.
- Isotropic and uniform grain growth. Particularly, the velocity of the grain growth is assumed to be equal in (i) all grains and (ii) all directions.

- Grain growth in a direction stops as two grain boundaries contact each other, i.e. there is no grain overlapping. The growth process stops, if there is no further grain growth in any direction in any grain.

- There are no voids, i.e. the entire volume is populated by grains.

It was found by Kumar and Kurtz (1994, 1995) that several statistical parameters are closely related for the artificial microstructure and a real aluminum titanium alloy (Williams and Smith, 1952) (Tab. 3.1). Additionally, the minimum surface energy

	$\bar{n}_e$	$\bar{n}_f$	$\bar{n}_c$	$\bar{\varphi}$	$\bar{\beta}$
Voronoi cell	5.228	15.536	27.086	120°	111.11°
Aluminum-Tin alloy grains (Williams and Smith, 1952)	5.020	12.480	20.880	-	-

($\bar{n}_e$ av. edges per face, $\bar{n}_f$ av. faces per cell,
$\bar{n}_c$ av. corners per cell, $\bar{\varphi}$ av. dihedral angle, $\bar{\beta}$ av. bond angle)

Table 3.1: Comparison of properties of the model and the real microstructure (Kumar and Kurtz, 1994)

theory requires $\bar{\varphi}$=120°and $\bar{\beta}$ ≈109° 28' which is close to the average values given in tab. 3.1. It is, hence, assumed that the model structure provides a sufficiently good approximation of the microstructure of polycrystalline aggregates.

Although these results suggest that the Voronoi tessellation is an adequate approximation for polycrystals, it has previously been stated (e.g., by Döbrich et al., 2004) that the variability of the grain size is highly underestimated, while the number of first neighbors is overestimated. Both can be influenced by the choice of generator points (see next section). Despite the stated over- and underestimations, a significant variability in the grain size, shape and neighbor relationship, can still be reproduced while providing a well defined geometric description of the aggregate.

3.3.1.1 Modification of the generator points

In order to overcome the shortcomings found in the comparison of the artificial microstructures and real materials, a modified set of Voronoi generators can be used. Two popular methods for the generation of such modified point sets are the hardcore Voronoi tessellation or the centroidal Voronoi tessellation (Du et al., 1999; Lautensack et al., 2008). The former is obtained by prescribing a minimum distance between two distinct Voronoi generators, as illustrated in Fig. 3.8. A physical motivation for the point constraint is given by the minimum size of stable nuclei in a melt.

The generation of a centroidal tessellation is more involved. It requires that the center of mass of each cell coincides with the vertex generating the respective cell (Fig 3.9). Such tessellations can be generated iteratively by starting from an unconstrained point set. The associated Voronoi diagram is computed and the centers of mass of all cells are evaluated.

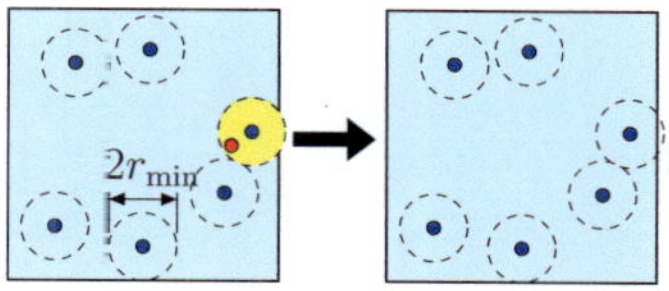

Figure 3.8: Hardcore condition for the point seed(cf. Fritzen et al., 2009)

If the distance between each Voronoi generator and the center of mass of associated cell is smaller than a prescribed tolerance, then the iteration is stopped and the centers of mass are taken as the constraint generator set. If not, a new tessellation is computed using the centers of mass as the new iterates of the Voronoi generators. The proposed algorithm was found to converge within few iterations.

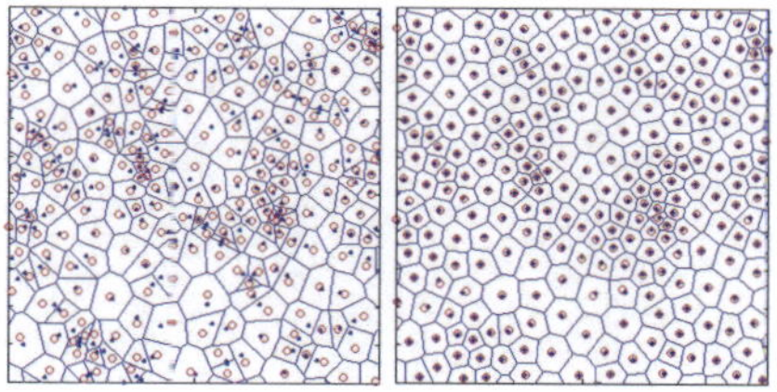

Figure 3.9: Comparison of a random Voronoi (left) and the resulting centroidal Voronoi tessellation (right) in 2d after 5 iterations (250 Voronoi generators); shown are the centers of mass and the Voronoi vertices (cf. Fritzen et al., 2009)

The cells generated using either of the two proposed point constraints are more regular, i.e. bad aspect ratios or small cells (Fig. 3.10) can be prevented. The regularity of the grain structure is represented by almost isotropic n-point statistics even for aggregates containing few grains. Consequently, the shape of the cells can be controlled by modification of the point seed in order to investigate the effect of variations in higher statistical moments onto the results of numerical simulations. Additionally, a wide literature on the statistical properties of the Voronoi tessellation exists, see for instance Aurenhammer (1991); Kumar and Kurtz (1995); Lautensack et al. (2008) and references therein. The aspect of the incorporation of the periodicity and its significance for computational homogenization was mentioned first by Decker and Jeulin (2000) and used in many subsequent works, e.g. by Barbe et al. (2001); Fritzen et al. (2009).

When restrictions on the generator points are imposed, it is important to also consider these constraints in the periodization algorithm. For example, the minimum distance prescribed by a hardcore constraint has to be computed based on the total point set containing the copied vertices. In case of the centroidal type of tessellation, only the centers of mass found inside the unit cell are considered. Based on these the 26 translated copies are created to provide the full set of all generators of the periodic tessellation.

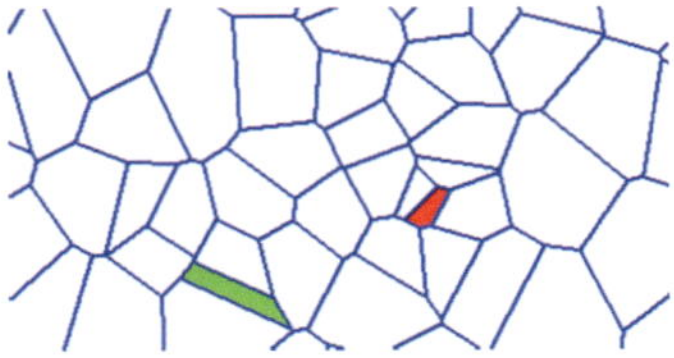

Figure 3.10: 2d Voronoi cell with high aspect ratio (green) and small Voronoi cell (red) in a 2d Voronoi tessellation (cf. Fritzen et al., 2009)

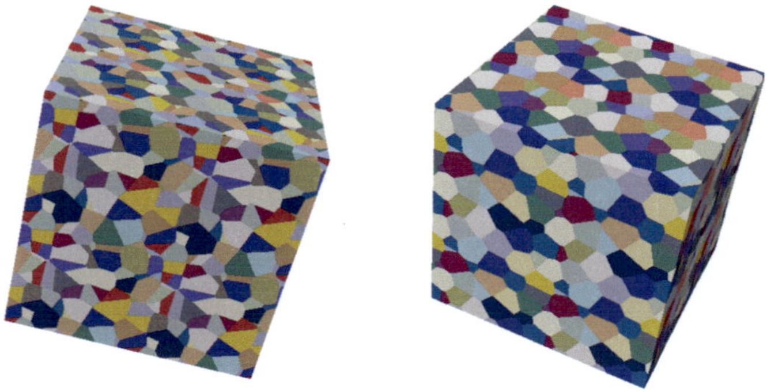

Figure 3.11: Example of a periodic, unconstrained Voronoi tessellation (left) and a periodic, centroidal tessellation (right) (cf. Fritzen et al., 2009)

In the following, unconstrained point sets and point sets subject to a hardcore condition are examined in the context of polycrystalline cubic aggregates. The focus is on periodic microstructures which have favorable numerical properties in the context of computational homogenization. Examples of crystal aggregates created based on the Voronoi tessellation and the centroidal Voronoi tessellation are shown in Fig. 3.11. The grains in the structure based on the centroidal tessellation are almost spherical while the ones in the standard tessellation often have certain preferred orientations. While both are statistically isotropic for infinite aggregates, the centroidal type of tessellation produces a closer approximation of the expected isotropy than the standard tessellation, for the same number of generator points $N_{\rm g}$.

3.3.1.2 Consideration of anisotropic grain morphology

Although the overall statistics of the Voronoi tessellation are isotropic when considering a large number of grains, many real world materials show a grain shape that is highly anisotropic due to the underlying manufacturing process. For example, the morphology of a sheet metal having undergone a rolling process has a pronounced directional anisotropy of the grain shape. The following algorithm is applied to the initially statistically isotropic aggregate in order to create model microstructures accounting for a class of morphological anisotropy:

- Generate a point seed using a Poisson process. [Optional: apply point constraints]
- Compute the Voronoi tessellation.
- Transform all corner points $\boldsymbol{x}_i$ of the cells of the tessellation according to

$$\tilde{\boldsymbol{x}}_i = \bar{\boldsymbol{F}}\boldsymbol{x}_i, \quad \bar{\boldsymbol{F}} = \begin{pmatrix} \lambda_1 & 0 & 0 \\ 0 & \lambda_2 & 0 \\ 0 & 0 & (\lambda_1\lambda_2)^{-1} \end{pmatrix}_{\boldsymbol{e}_k \otimes \boldsymbol{e}_l}, \tag{3.18}$$

  where $\lambda_1, \lambda_2 > 0$ describe the macroscopic deformation that the material is subjected to in the production process. A rolling process can, for instance, be approximated by

$$\bar{\boldsymbol{F}} = \begin{pmatrix} \lambda & 0 & 0 \\ 0 & \lambda^{-1} & 0 \\ 0 & 0 & 1 \end{pmatrix}_{\boldsymbol{e}_i \otimes \boldsymbol{e}_j} \qquad (\lambda > 1). \tag{3.19}$$

- Use the resulting geometry for further processing, e.g., for the generation of a spatial discretization (see section 3.3.1.3).

The proposed algorithm preserves the periodicity of the geometry and the volume fraction of the individual grains, i.e., the one-point auto-correlation function remains constant. Consequently, it allows to study the effect of higher order statistics at constant one point statistics (i.e., volume fractions) and with identical grain-to-grain neighborhood (see also Böhlke et al. (2009); Jöchen et al. (2010)). Examples of a periodic Voronoi tessellation with and without point constraint and a morphological anisotropy of the form (3.19) are shown in Fig. 3.12.

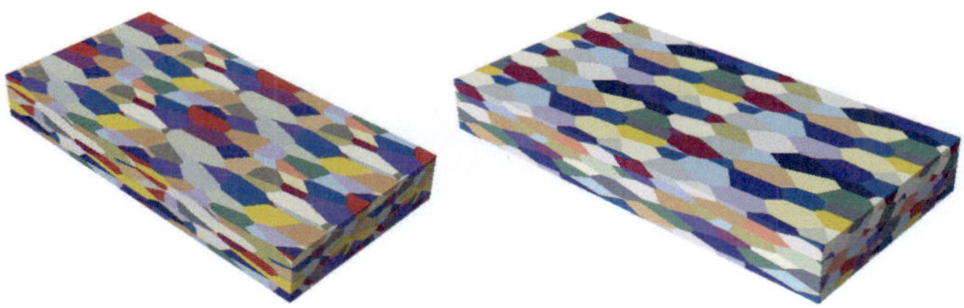

Figure 3.12: Example of a periodic, unconstrained Voronoi tessellation (left) and a periodic, centroidal tessellation (right) subjected to a rolling morphology ($\lambda = 4$)

3.3.1.3 Periodic spatial discretization of crystal aggregates

In the following a detailed description of a hierarchical mesh generation procedure used for the spatial discretization of Voronoi tessellations is presented. An overview of the algorithm is shown in Fig. 3.13. The algorithm is defined by a hierarchy of discretizations with increasing spatial dimension. In each mesh generation step the spatial dimension of the mesh increases, with the mesh of the previous step defining the constraint boundary for the next step, i.e. only the *interior mesh* is generated in each step. Starting from a 0-dimensional mesh (the point pair defining an edge of the tessellation), the *interior* mesh consists of additional co-linear points defining line segments. In the next level planar facets bounded by the previously computed line segments are meshed, forming the constraint boundary for the generation of the volume mesh. The proposed mesh generation procedure is implemented into a C++ program, which allows for the use of external libraries and superior performance.

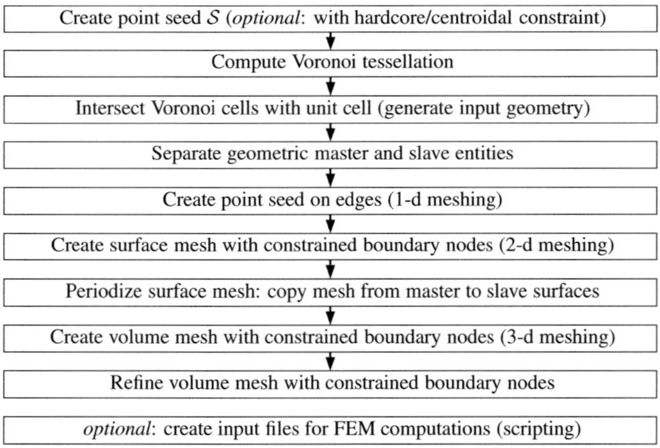

Figure 3.13: Hierarchical structure of the meshing algorithm (cf. Fritzen et al., 2009)

Before the actual mesh generation procedure can be applied, it is necessary to determine the geometry of the tessellation in a first step. Therefore, the point seed $\mathcal{S}$ is created by a Poisson process. By N_g the number of nuclei is denoted. In a Poisson process each point $\boldsymbol{p}_i$ ($i = 1, \ldots, N_g$) is assigned an x-, y- and z-coordinate via random variables X_i, Y_i, Z_i which are uniformly distributed around the origin at (0,0,0) with

$$X_i \sim \mathcal{U}([-\tfrac{w}{2}, \tfrac{w}{2}]),\ Y_i \sim \mathcal{U}([-\tfrac{d}{2}, \tfrac{d}{2}]),\ Z_i \sim \mathcal{U}([-\tfrac{h}{2}, \tfrac{h}{2}]),$$

where w, d and h are the width, depth and height of the cuboid unit cell, respectively. For the Hardcore Voronoi tessellation all points $\boldsymbol{p}_j$ with

$$\|\boldsymbol{p}_i - \boldsymbol{p}_j\|_2 \leq r_{\min} \qquad (j > i) \tag{3.20}$$

are deleted from the point seed $\mathcal{S}$ (Fig. 3.8) and new random points are added. This procedure is repeated until the number of points satisfying the prescribed hardcore radius is obtained. The point seed is then copied 26 times to accommodate the periodicity constraint.

The computation of the Voronoi tessellation itself is carried out by using the `QHull` package (Barber et al., 1996) which is fast and robust. The output consists of a set of vertices $\mathcal{P}$ and a list of corner vertices for each Voronoi cell. It is noteworthy that a total of 27 N_g Voronoi cells is obtained because of the periodization procedure. Most of the resulting cells are found exterior of the unit cell. Other cells intersect the boundary of the cell. The removal of these exterior cells and the intersection of the Voronoi cells with the bounding box is described in the following.

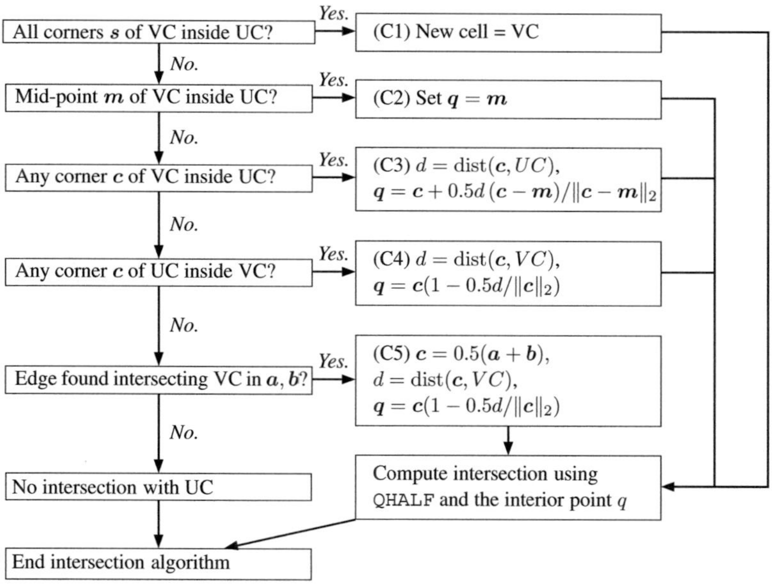

Figure 3.14: Algorithm for intersection of a Voronoi cell with the unit cell (cf. Fritzen et al., 2009)

The tool `qconvex` (part of `QHull`) provides a representation of the Voronoi cell in terms of its bounding hyperplanes (more precisely, of the half-spaces whose intersection is defining the domain of the cell). By adding the six half-spaces defining the bounding box to this list, the intersection can be computed using `qhalf` which is also part of `QHull`. The crux in this procedure is the identification of a point which lies *truly* inside (i.e. not on the boundary) of both, the bounding box and the Voronoi cell. This point is required for `qhalf`. The identification of the interior point is pursued based on the algorithm shown in Fig. 3.15. A geometric interpretation of the cases C1-C5 is presented in Fig. 3.15, where the

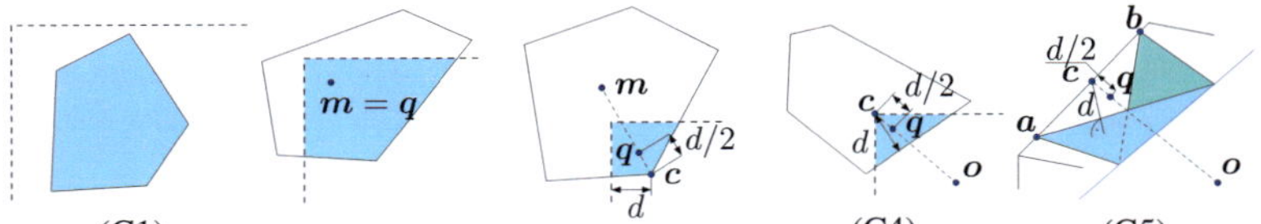

(C1) (C2) (C3) (C4) (C5)

Figure 3.15: Graphical interpretation of the intersection cases (C1)-(C5) (cf. Fritzen et al., 2009)

two-dimensional representation has been chosen for C1-C4 for the sake of simplicity. The algorithm exploits the convexity property of both the unit cell and the individual Voronoi cells to find a *truly* interior point. The convexity of the two domains further ensures the convexity of their intersection, i.e. each cell fragment is convex.

The output of `qhalf` is merged to the existing points and facets. When merging the data of the individual cells, all (local) point indices have to be reindexed appropriately. A delicate topic is the removal of duplicate entities which is necessary to avoid errors in the later stages of the mesh generation procedure. In this step, the fact that the discrete geometry is entirely described based on points and point indices can be used efficiently. As a consequence, it suffices to remove duplicate points, i.e., all point pairs showing a distance smaller than a given numerical precision δ_p are merged into one point. Since point indices are altered, all data structures containing point indices need to be reindexed after the duplicate point removal. Then all duplicate facets can be eliminated from the input data. It is worth mentioning that each facet which is not found on the boundary of the unit cell occurs twice with reverse orientation, since two neighboring grains share one adjacent facet with inversely oriented outer normal vector.

In order to be able to apply periodic fluctuation boundary conditions, the periodicity of the mesh topology is of outmost interest. In order to enforce the periodicity, three master and slave surfaces are defined on the unit cell. The master regions

`Front`: $\{x = \frac{w}{2}\}$, `Right`: $\{y = \frac{d}{2}\}$, `Top`: $\{z = \frac{h}{2}\}$

were decided on. The corresponding slave surfaces `Back`, `Left`, `Bottom` are defined analogously with the respective negative axial offset. All facets found on these slave surfaces are removed since they are translated copies of their respective counterpart on the master surfaces. A periodic mesh is characterized by periodic continuation of all mesh entities on the pairings of master and slave surfaces. The creation of such a conforming surface mesh exploits the hierarchical topology of the model microstructure, i.e. the surface mesh is the interface between two neighboring cells and the interface between two neighboring facets is one common edge. By prescribing the nodes on the edge interfaces it is possible to create compatible surface meshes on all facets individually. Based on this fact, the creation of the surface mesh is divided into the following steps:

- Edge splitting:
  - Creation of an edge list (each edge is characterized by 2 point indices);
  - Split edges with length $l > \hat{l}$ into $n_s = l/\hat{l} + 1$ segments of length $l_s = l/n_s \leq \hat{l}$, where the maximum length $\hat{l}$ is computed from the prescribed volume $\hat{v} = \frac{wdh}{N_{el}}$:

    $$\hat{l} = \sqrt[3]{\hat{v}/0.09},$$

    where the factor 0.09 is derived from the volume v_r of a regular tetrahedron

    $$\hat{v} \approx 0.75 v_r = 0.75 \tfrac{\sqrt{2}}{12} \hat{l}^3 \approx 0.09 \hat{l}^3;$$

- For each facet (see also Fig. 3.16):
  - Transform all points found on bounding edge segments from 3d into 2d;
  - Mesh facet using the `Triangle` library with constraint boundary nodes;
  - Transform 2d points back to 3d;

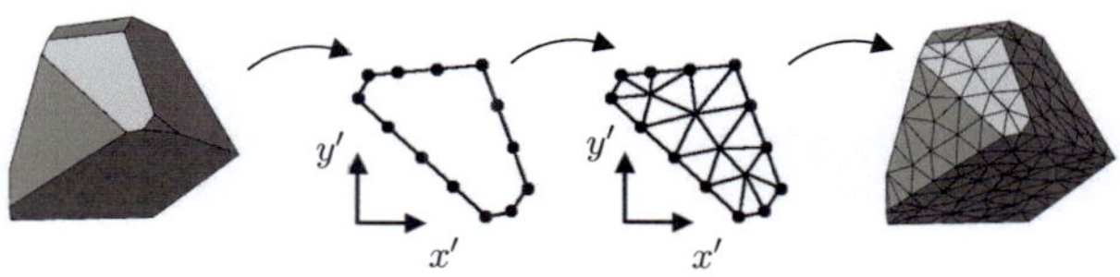

Figure 3.16: Creation of the surface mesh on a single facet

  - Merge new points and triangles to the vertex list and to the surface mesh respectivley;

- Copy mesh from the master to the slave surfaces (Fig. 3.17);

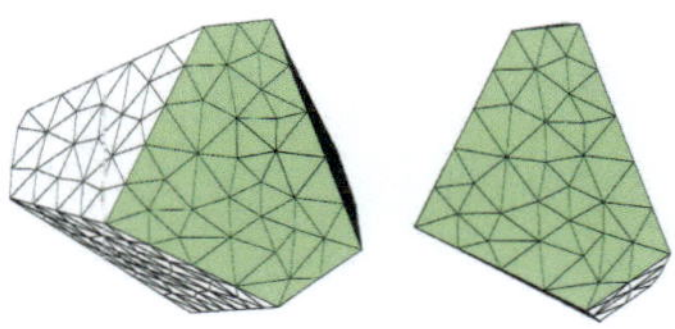

Figure 3.17: Copying of nodes and boundary triangles from a master (left) to the corresponding slave (right) surface

- Remove duplicate points and reindex surface triangulation.

By copying the triangular surface mesh from the master onto the slave surfaces, the periodicity constraint is automatically satisfied. At this point, the regular edge splitting algorithm plays a crucial role, since edges on opposing sides of the unit cell automatically share the same segments by construction. Moreover, the surface mesh created by `Triangle` (Shewchuk, 1996) is of high quality if the input angles are not too small. The latter is the case for most facets. Unfortunately, small input angles may occur. This effect is more pronounced on the boundary, due to the intersection of the Voronoi cells with the unit cell. The removal of these would result in a modified topology and is, therefore, omitted.

The volume mesh is created based on the periodic surface mesh using the `Tetgen` library (`http://tetgen.berlios.de/index.html`; Si and Gaertner (2005)). Additional parameters are used to assign the material number to each cell. The meshing is carried out in two stages to increase robustness:

[S1] Mesh with constraint boundary nodes
$\longrightarrow$ low quality tetrahedral mesh;

[S2] Remesh (constraint boundary nodes) with prescribed maximum tetrahedral volume
$\longrightarrow$ high quality tetrahedral mesh.

3.3.1.4 Example meshes

In this section, some example meshes created using the previously outlined algorithm are presented. The quality of the generated elements is briefly discussed. The surface triangulation on the boundary of a unit cell consisting of 200 grains is shown in Fig. 3.18. This discretization consists of a total of 234'946 nodes and 171'247 tetrahedra with quadratic ansatz functions. Fig. 3.19 represent a cut through the cell to illustrate the homogeneous mesh quality in the interior of the domain.

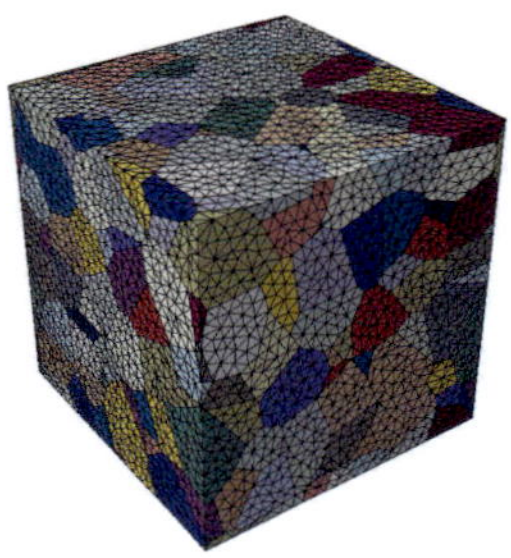

Figure 3.18: Outer surface mesh

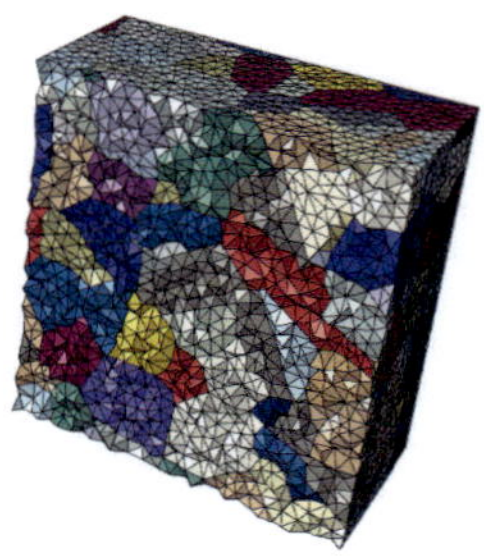

Figure 3.19: Cut through the unit cell

The assignment of material properties is conducted via element sets. The existence of these allows for convenient pre- and post-processing operations in the context of finite element computations. All materials can be viewed individually as illustrated in Fig. 3.20. The

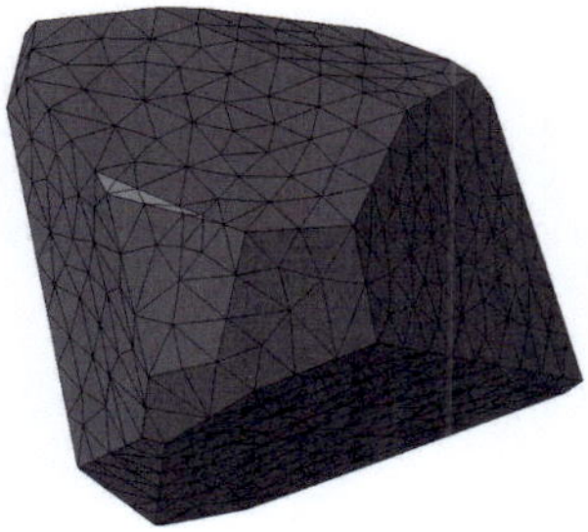

Figure 3.20: Single grain

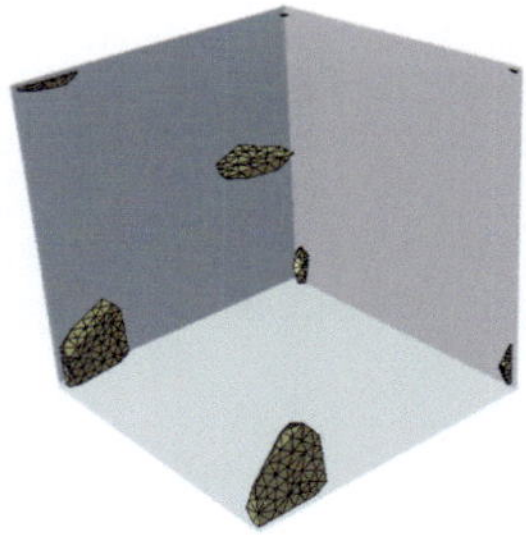

Figure 3.21: Grain with periodic continuation

identification of the periodic continuation of grains on opposing sides of the unit cell is also simplified by this technique. The periodic continuation of a grain intersecting the boundary of the unit cell is shown in Fig. 3.21.

The proposed algorithm allows the comparison of the constitutive response of the identical geometry using varying mesh densities. Four different mesh densities for an aggregate containing 200 particles are shown in Fig. 3.22 for comparison. The latter is an important issue when investigating convergence with respect to the mesh size. In particular, the same *geometry* can be used for a variety of different *discretizations* without having to account for possible volumetric defects that can occur in both, image based mesh generation techniques (e.g., based on the programs AVIZO or ScanIP/ScanFE) and the multi-phase finite element method (see, e.g., Zhodi and Wriggers, 2005).

Figure 3.22: Different mesh densities: 58'268 (41'203), 234'946 (171'247), 929'538 (689'825), 1'753'990 (1'309'083) nodes (tetrahedra) (left to right)

The shape factors* of the different discretization levels are 0.5190, 0.6342, 0.6541 and 0.6570, respectively. Remarkably, the creation of the finest mesh containing a total of 1.754 million nodes took only two minutes seconds on an ultra-mobile notebook computer using only a single core of the CPU. Memory requirements are negligible for the proposed method

*Definition of the shape factor: $\bar{\eta} = \frac{v}{v_{\text{opt}}}$ with the volume v_{opt} of an equilateral tetrahedron with the same circumradius.

with respect to current standards. These values highlight the efficiency of the chosen implementation. The number of performed clipping operations for the intersection of the Voronoi cells the unit cell is exemplified by the large number of 388 cell fragments for 200 generator points, i.e. 188 additional cell fragments at the boundary were generated. The proposed discretization technique is used for the computational homogenization of the elastic properties of polycrystals with cubic symmetry in section 6.1.1 (see also Fritzen et al., 2009).

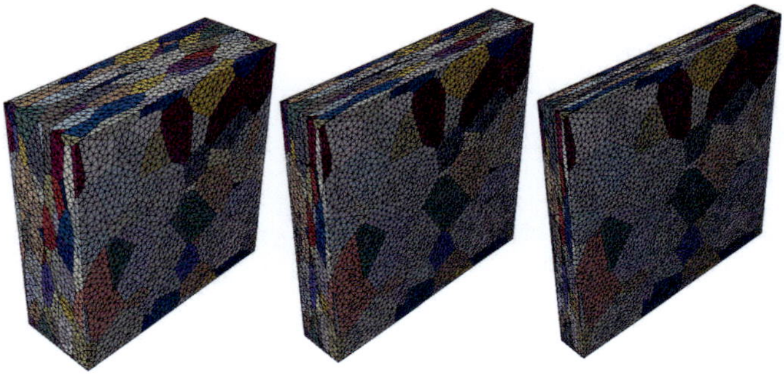

Figure 3.23: Microstructures with oblate grain shape and grain aspect ratios of $2\sqrt{2}, 3\sqrt{3}$ and 8 (left to right)

Examples of anisotropic meshes with oblate grain morphology and varying degree of anisotropy are shown in Fig. 3.23. Discretizations of aggregates with elongated grains are presented in Fig. 3.24 for different anisotropy parameters. All shown discretizations are composed of 200'000 to 600'000 nodes and 140'000 to 450'000 tetrahedra. The underlying periodic Voronoi tessellation was generated from a set of 200 generator points plus the obligatory 5'200 translated copies of these points. The algorithm for the modified tessellations was used to investigate the importance of the grain morphology in section 6.1.2 (see also Böhlke et al., 2009; Jöchen et al., 2010).

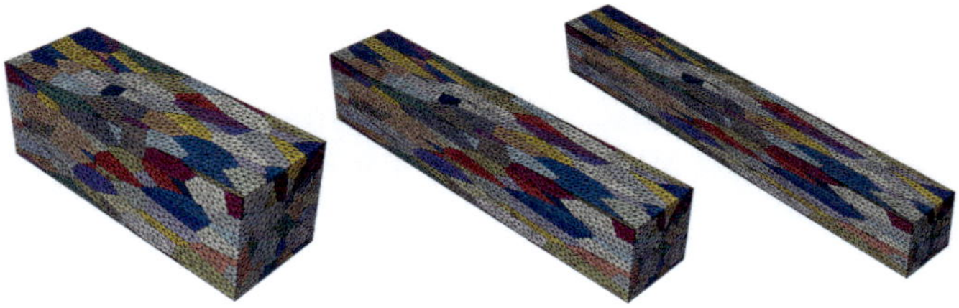

Figure 3.24: Microstructures with elongated grain shape and mean grain aspect ratios of $2\sqrt{2}, 3\sqrt{3}$ and 8 (left to right)

3.3.2 Particle reinforced composites

Modern composite materials often consist of one phase containing one or several other constituents of particulate morphology. Schematic examples of the observed particle shapes are shown in Fig. 3.25. In the following the phase containing the inclusions is referred to as matrix phase and the particles are called reinforcement. Typical examples for materials with particulate reinforcements are various metal ceramic composites (often: MMC = metal matrix composite) (e.g., Miserez et al., 2004; Ganesh and Chawla, 2005; Chawla and Chawla, 2006) or particle reinforced rubber-like materials. The presented algorithm was first presented in Fritzen and Böhlke (2011c) and later modified to allow for anisotropic particle shapes in Fritzen and Böhlke (2011b).

Figure 3.25: Examples of materials with particle reinforced matrix: statistically isotropic polyhedral particles, oblate particles, elongated particles (left to right)

Based on the periodic Voronoi tessellation introduced in the previous section in detail, a suitable modification to attain a particulate microstructure is found by erosion of the cells to obtain the particles and to fill the intermediate space with the matrix material. Therefore, two natural approaches can be pursued: (i) uniform centroidal scaling (modification of the cell's corner vertices) or (ii) modification of the offsets of half-space defining the cell. The latter is a process very similar to the erosion algorithm commonly used in image processing (see, e.g., Ragnemalm, 1992). It was decided on the second proposal for the following reasons:

- The nearest neighbor distance of the particles is homogeneously distributed.
- Arbitrary particle volume fractions are possible.
- The number of faces of small particles is (usually) smaller than of large particles since small faces are removed during the erosion of the cells. This corresponds to experimental observations, where smaller particles often show a smaller number of bounding facets. This is partially due to the smaller particles resulting from fracture of originally larger particles.

A similar algorithm was proposed by Christoffersen (1983) for geomaterials and for the use with semi-analytic homogenization methods. The following representations are different in the sense that a periodic structure is created and that a three-dimensional description is attained. Using the same geometric manipulation described in section 3.3.1.2 the morphology of the Voronoi tessellation can be modified before the erosion cells. Particles created based on such modified tessellations exhibit an morphological anisotropy as previously reported for the grains in the artificial crystal aggregates.

3.3.2.1 Mesh generation algorithm for particulate materials

Similar to the algorithm presented in section 3.3.1.3 for crystal aggregates, a hierarchical mesh generation technique was developed to allow for fully periodic mesh topologies. The disjoint particles and the remaining non-convex matrix material require a modified algorithm for the creation of a quality surface triangulation and, based on the latter, a high quality volume mesh. A detailed description of the algorithm is presented by Fritzen and Böhlke (2011c), where additional algorithmic informations can be found. The main steps of the algorithm for the creation of a model microstructure with a particle volume fraction c are:

[P1] Create the random point seed and the corresponding periodic Voronoi tessellation (*optional:* hardcore condition, centroidal tessellation; anisotropic modification; see Fig. 3.26).

Figure 3.26: Comparison of the particle shape obtained from an unconstraint Voronoi tessellation (left) and a centroidal Voronoi tessellation (right) (50 Voronoi generators, 30% particle volume fraction)

[P2] For each cell, the half-space representation $\mathcal{H} = \{\boldsymbol{n}_i, d_i\}_{i=1,\dots,N_{\text{facets}}}$ is computed for all cells of the tessellation. Then the points $\boldsymbol{x}$ found within the cell $\mathcal{C}$ defined by $\mathcal{H}$ satisfy the inequality

$$\boldsymbol{n}_i \cdot \boldsymbol{x} + d_i \leq 0 \qquad \forall \boldsymbol{x} \in \mathcal{C}. \tag{3.21}$$

The offsets d_i of the half-spaces are modified according to

$$d_i \rightarrow d_i + \delta, \tag{3.22}$$

where $\delta > 0$ is half of the distance w between neighboring particles (Fig. 3.27).

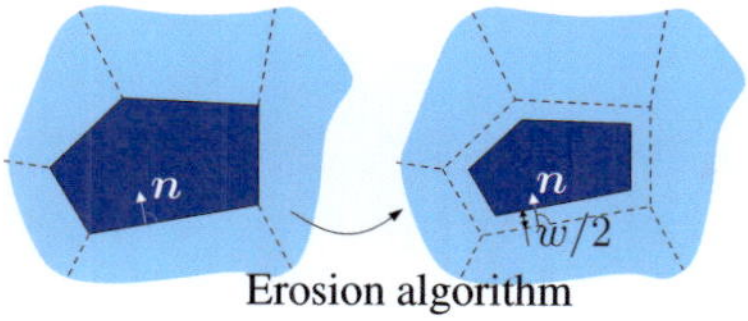

Figure 3.27: Schematic representation of the erosion of a single two-dimensional Voronoi cell

[P3] The volume fraction is computed and **[P2]** is repeated with adjusted values of δ, until the prescribed particle fraction $c \in (0, 1)$ is obtained.*

[P4] For each of the resulting particles the intersection with the cuboidal unit cell is computed.

[P5] A conforming edge point seed is created by splitting all edges defining the geometry to satisfy a maximum length criterion (Fig. 3.28).

[P6] A surface triangulation of all particle facets is computed (Fig. 3.28; see also 3.3.1.3).

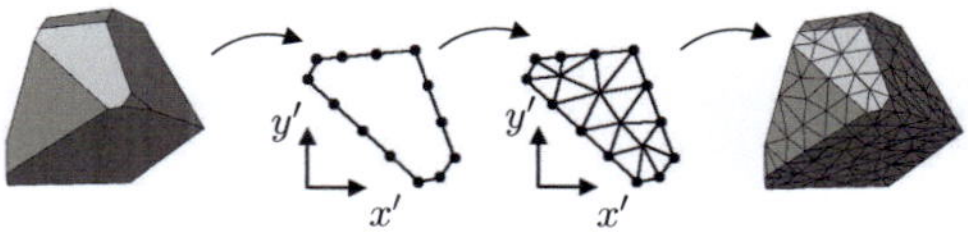

Figure 3.28: Creation of the surface mesh on a single facet of a particle

[P7] Create a constraint surface triangulation on the six surfaces of the cuboid with holes for the particle facets located on the boundary (Fig. 3.29). The mesh on the slave surfaces is obtained by copying the mesh of the master surfaces.

[P8] Assemble all surface triangulations to the overall surface mesh of the particulate microstructure and create a quality volume mesh using `tetgen` (Si and Gaertner, 2005).

[P9] Post-process generated tessellation to create material information, element and node sets and input files for further processing in finite element computations.

It is noteworthy that the presented algorithm is sufficiently robust to create meshes for tessellations containing from 5% up to 90% of particles. Many existing algorithms based on spheres or special polyhedra (e.g., Flaquer et al., 2007; Nogales and Böhm, 2008) can only be used for moderate volume fractions of up to 30% when using the random sequential addition method. In order to achieve higher volume fractions, techniques such as simulated annealing or random sequential adsortion are often required. These techniques influence the

*A detailed algorithm for the iteration including a highly usable initial guess can be found in Fritzen and Böhlke (2011c).

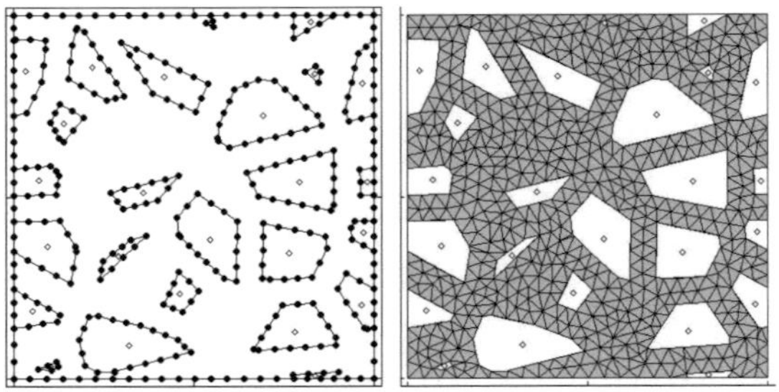

Figure 3.29: Common edge seed of the particles and the matrix on the exterior faces of the cuboidal cell (left) and constraint surface mesh on the boundary for the matrix material only (right)

statistical properties of the microstructure and no immediate conclusions with respect to statistical isotropy are possible *a priori*. Additionally, particle fractions above approximately 60% cannot be realized with most methods.

3.3.2.2 Example meshes

A selection of example meshes is used to illustrate the robustness and wide range of operation of the proposed algorithm. First, variations of the number of particles and of the particle volume fraction are considered in Fig. 3.30. Between 10 and 100 particles are considered for volume fractions ranging between 10 and 60 %. The mesh density for the shown discretizations is chosen such that each mesh consists of ≈100'000 nodes. The computational time for the creation of the meshes ranged between 30 and 60 seconds. Hence, the computational efficiency of the method can be considered as very good. An appealing feature of the

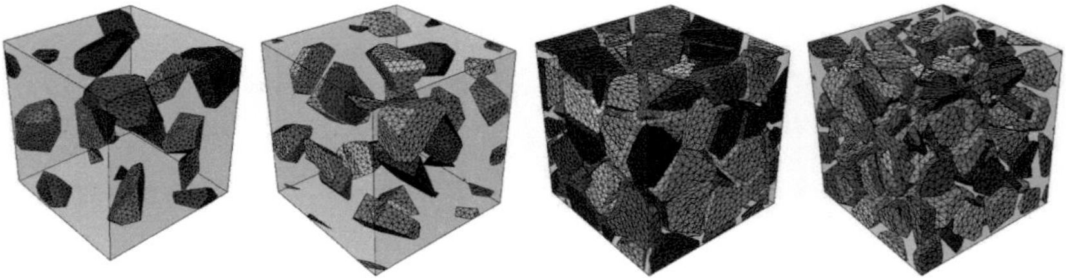

Figure 3.30: Meshes containing 10, 20, 50 and 100 particles (10%, 10%, 60%, 30% particle volume fraction)

algorithm is found in the fact of fully adjustable mesh density. This allows for quick calculations on coarse grids to attain a qualitative impression of the thermo-mechanical response of particulate structures. The precision of the prediction can then be refined at will. Mesh density studies can be performed to investigate the convergence behavior. Fig. 3.31 shows different discretization levels of the same underlying geometry. With respect to the particle

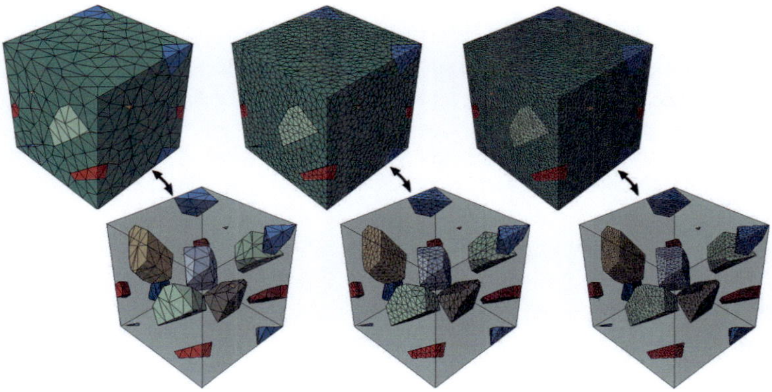

Figure 3.31: Different discretization levels 6502 (left), 40531 (middle) and 363808 (right) nodes for 10 particles (10 % vol. frac.) (see Fritzen and Böhlke, 2011c, for details)

volume fraction, the algorithm is found to produce high quality discretizations even for high particle volume fractions such as 50% or 80% (Fig. 3.32). Real materials with high particle volume fractions and polygonal particle shape are investigated, e.g., by Miserez et al. (2004) in terms of B_4C or Al_2O_3 reinforcements in an aluminum matrix material.

Figure 3.32: Examples for 50 % (left) and 80 % (right) particles ($\approx$200000 nodes, 200 Voronoi generators)

With respect to anisotropic modifications it is referred to section 6.5, where the presented algorithm is combined with the affine transformation presented for the polycrystalline aggregates in section 3.3.1.2. After the erosion of the cell walls of the initially isotropic Voronoi cells, the particles are subjected to an affine transformation which leads to an anisotropic particle shape. The generated mesh is still of high quality since the geometry is modified before the mesh generation. Further details on the algorithm are presented in Fritzen and Böhlke (2011b) and example discretizations are shown in section 6.5.

3.4 Porous materials

3.4.1 Microstructure generation

The nucleation and growth of pores in polycrystalline metals during plastic deformation is the main mechanism leading to ductile failure. Due to the large industrial interest in the understanding and, moreover, prediction of ductile failure, the literature on the topic is exhaustive. A milestone in the prediction of the pressure sensitivity of the effective yield stress of porous metals was the early work of Gurson (1977) in which the relation

$$\bar{\sigma}_{\mathrm{vM}} = \sigma_{\mathrm{F}} \sqrt{1 + c^2 - 2c \cosh\left(\frac{3\bar{\sigma}_{\mathrm{vM}}}{2\bar{\sigma}^\circ}\right)} \tag{3.23}$$

for the pore volume fraction c, the applied hydrostatic stress $\bar{\sigma}^\circ$ and the applied von Mises equivalent stress $\bar{\sigma}_{\mathrm{vM}}$ was derived. The yield stress σ_{F} is assumed constant.

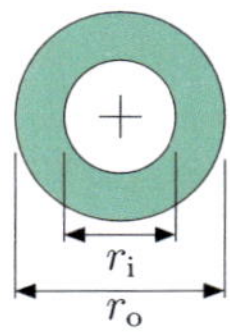

Figure 3.33: Reference cell used in the derivation of the Gurson model

The relation (3.23) was obtained using the reference cell Fig. 3.33, where the boundary values of the displacement are given by the applied (macroscopic) strain. The cell consists of a hollow sphere where the inner radius r_{i} and the outer radius r_{o} are related by the pore volume fraction c according to

$$r_{\mathrm{o}} = r_{\mathrm{i}} c^{-1/3}. \tag{3.24}$$

The cell in Fig. 3.33 is suitable for semi-analytic examinations and a wide literature exists with the aim of increasing the quality of the prediction (3.23). Some notable contributions to the topic are due to Tvergaard (1981); Ponte-Castañeda (1991); Garajeu et al. (2000); Garajeu and Suquet (2007); Monchiet et al. (2007). In essence, all these methods refer to reference cells similar or identical to the one used in the derivation of Gurson's model. However, different model assumptions, e.g., in terms of Eshelby type trial velocity fields in Monchiet et al. (2007), are used to determine the effective response. In order to evaluate the efficiency of the different approaches with respect to more realistic porous microstructures, ensembles containing impenetrable spherical pores of constant diameter are created randomly. A Poisson point process generates the centers of the spheres, while the radius is computed from the desired pore volume fraction c and the number of spheres in the unit cell N via

$$r = \sqrt[3]{\frac{3|\Omega|}{4\pi N}}. \tag{3.25}$$

Each generated point is copied 27 times in order to generate a periodic microstructure (see also 3.3.1), after a collision detection algorithm confirms that the newly added sphere does not intersect with existing spheres. For reasons associated with the later discretization a minimum distance of 2.5% of the sphere radius r is enforced between neighboring pores. Such an assemblage is also referred to as hardcore model of spheres in the literature (e.g. Torquato, 2002). Using the described random sequential addition procedure, pore volume fractions of up to 30% can easily be attained. Above 30% it is asserted that some (or most) pores intersect and that the pores can no longer be treated with the described geometry. The limit volume fraction for spheres without repulsion distance was determined numerically by Cooper (1988). A range of 37-40% was found. Later the numerical predictions were refined and an asymptotic value of 38% was determined for a quasi-infinite number of spheres at a single size population. Examples of the porous material (drawn as green spheres) are provided in Fig. 3.34 for 10, 20 and 30% pore volume fraction and 50 voids each.

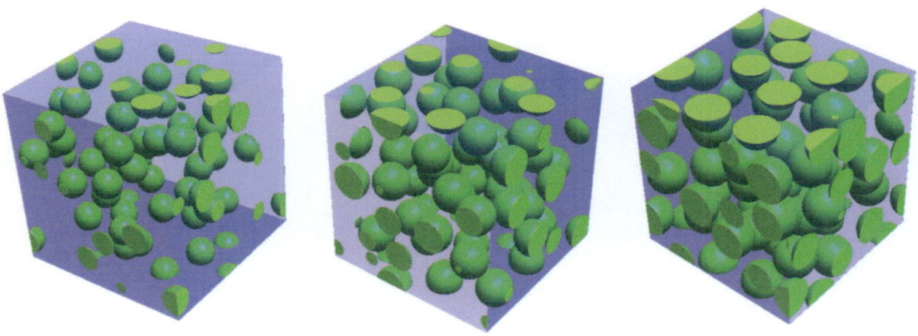

Figure 3.34: Random realizations of periodic unit cells representing porous metals

In order to illustrate the inhomogeneity of the ensembles, the k-nearest neighbor distance is drawn in Fig. 3.35 for 0.5, 1 and 5% pore fraction and in Fig. 3.36 for 10, 20 and 30% volume fraction. The results of five different statistical realizations are shown for each microstructure. Besides the average value $\bar{d}_k$ (lines) the minimal and maximal values are also presented (points). All values are given in units edge length of the reference cube (here: 1 μm). A quantification of the standard deviation is plotted in terms of the $\pm 3\sigma(\bar{d}_k)$ (dashed lines) confidence interval that should contain 99.7% of all points for a Gaussian distribution. It can be seen that the scattering between different realizations is more pronounced the smaller the pore volume fraction and the smaller k. Most notably, the nearest neighbor distance d_1 shows a wide spread. This is emphasized by massive violation of the provided confidence interval for $k \lessapprox 5$ to 7 (depending on the pore fraction). Generally, strongly varying local geometries result in heterogeneous solution fields under external loading. Evidently, the cell employed by Gurson and others can consider neither such interactions between neighboring pores nor their influence on the effective response of the material. In order to quantify the discrepancy of the analytical results with respect to these local fluctuations the artificial microstructures are used in numerical simulations.

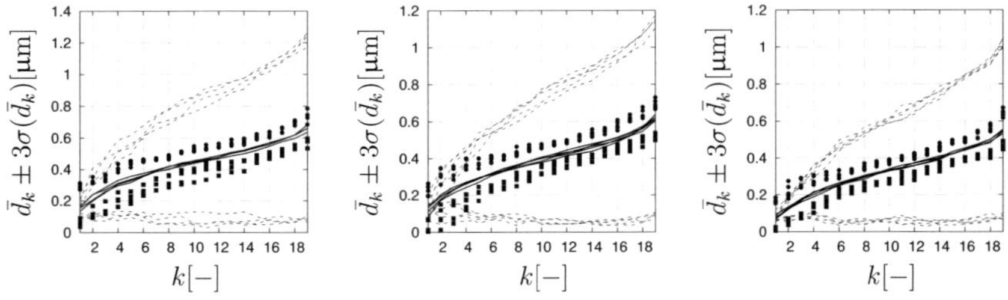

Figure 3.35: Average k-nearest distance of the pores for 0.1% (left), 1% (middle) and 5% pore volume fraction (right); 5 realizations with min./max. values, $k = 1, \ldots, 19$; dashed lines: 3σ confidence interval

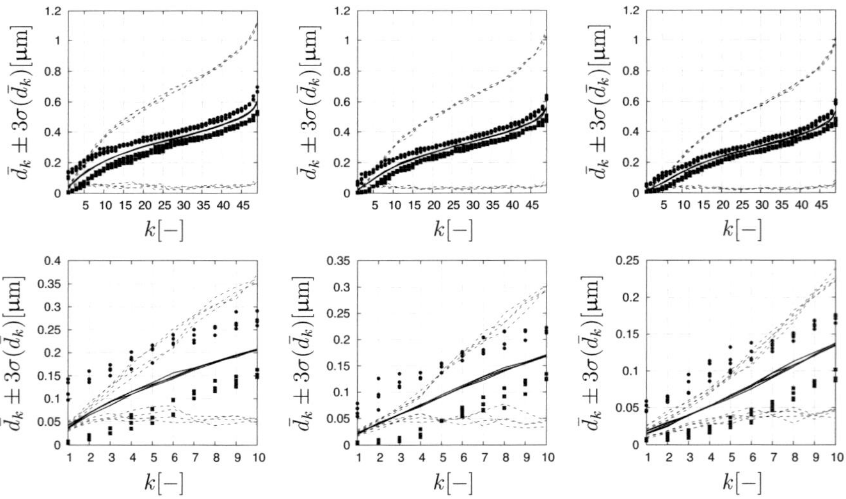

Figure 3.36: Average k-nearest distance of the pores for 10 (left), 20 (middle) and 30 % pore volume fraction (right); 5 realizations with min./max. values, $k = 1, \ldots, 49$ (top) and $k = 1, \ldots, 10$ (bottom); dashed lines: 3σ confidence interval

At low volume fractions (Fig. 3.35) with 20 voids considered, the minimal and maximal nearest neighbor distance shows distinct variations between different realizations of the microstructure. However, the average values for the different realizations are very close to each other (overlapping lines) and the standard deviation does not scatter too much. At the higher volume fractions (10-30%) and with more particles being considered, there is almost no observable scatter in $\bar{d}_k$ (Fig. 3.36).

In order to investigate the existence of pore clusters more closely, the algorithm presented in section 3.2.2.2 is used to detect and visualize pore ensembles. First, a unit cube showing 1% porosity (20 voids) is examined. The pore clusters determined from different values of δr are shown in Fig. 3.37. It is seen that small clusters consisting of two or three pores separated by thin walls exist. Larger values of δr reveal some local fluctuations in the pore density. For very large values of δr all particles tend to form a single cluster, i.e., such values are out of interest.

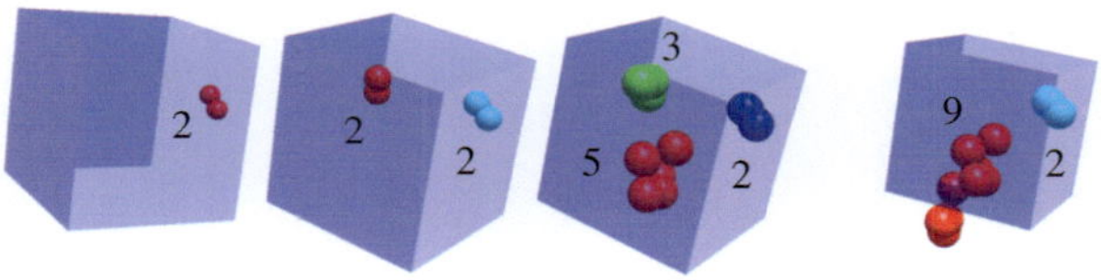

Figure 3.37: Detected pore clusters for $\delta r = 0.01, 0.03, 0.06, 0.07$ (left to right), $c = 0.1\%$, $N = 20$; numbers indicate cluster size (colours may vary)

The results obtained for only 20 pores are confirmed by a reference cell containing 100 pores at the same volume fraction (1%). It is found that for small values of δr, dispersed clusters are detected. As δr rises, the number of clusters increases and larger clusters are found, see Fig. 3.38.

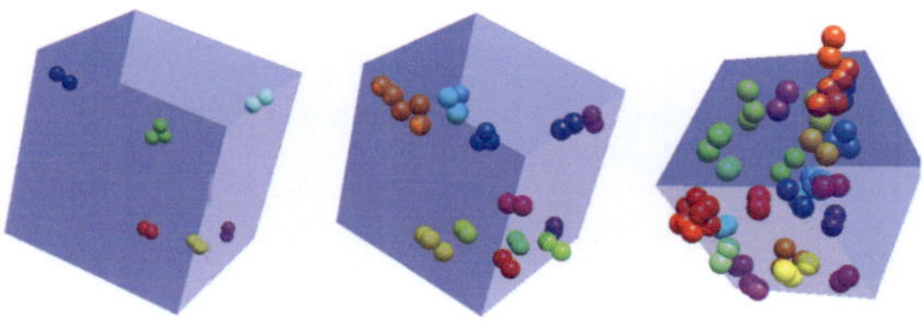

Figure 3.38: Detected pore clusters for $\delta r = 0.01, 0.02, 0.04$ (left to right), $c = 0.1\%$, $N = 100$ (colours of individual clusters may vary)

At pore volume fractions higher than 10%, the detection of clusters is of little use as the cluster topology changes rapidly with respect to δr. Generally, small micro-clusters consisting of two to four pores are often existing. None of the investigated microstructures shows massive pore agglomeration.

3.4.2 Mesh generation for porous materials

The discretization of the microstructures was performed using the mesh generation software `netgen` (Schöberl, 1997). A sufficiently fine discretization level was achieved by tuning the mesh generation parameters. Quadratic tetrahedral Lagrange elements are used. The periodicity was explicitly enforced in `netgen`. The generation of each mesh is performed based on 20 ($c = 0.1, 1, 2.5, 5\%$) and 50 voids ($c = 10, 20, 30\%$). Several successive refinement steps are performed to achieve optimal mesh quality. The number of nodes for the discretizations was approximately 100'000 ($c = 0.1\%$), 85'000 ($c = 1\%$), 85'000 ($c = 2.5\%$), 85'000 ($c = 0.1\%$), 65'000 ($c = 5\%$), 210'000 ($c = 10\%$), 120'000 ($c = 20\%$) and 130'000 ($c = 30\%$). The higher number of degrees of freedom for the smaller volume fractions is due to the increased curvature of the voids and the higher volume fraction of the metal. Examples of the discretizations are shown in Fig. 3.39. The generated discretizations show excellent mesh quality.

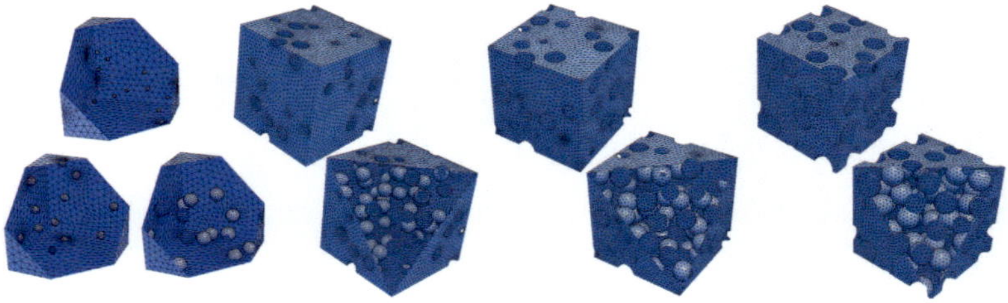

Figure 3.39: Finite element mesh for pore volume fractions 0.1%, 1% and 5%, 10%, 20% and 30% (left to right)

In addition to these mesh densities used in the statistical investigations, a set of coarser and finer discretizations was created for an ensemble of 10 voids for the use in a mesh density study. The number of nodes in the different discretization levels is shown in Tab. 3.2. Three of the discretizations (levels 1,3 and 6) are shown in Fig. 3.40. For details on the computational results see section 6.3.

refinement level	h_1	h_2	h_3	h_4	h_5	h_6
number of nodes	7'463	15'994	40'616	106'335	424'766	1'155'704
degrees of freedom	22'389	47'982	121'848	319'005	1274'298	3'467'112

Table 3.2: Mesh refinement levels for the mesh density study (10 voids, 20% vol. fraction)

Figure 3.40: Mesh of the refinement level h_1, h_3 and h_6

Chapter 4

Homogenization methods

4.1 Thermo-mechanical two-scale problems

Engineering applications usually consider real materials on a structural (or macroscopic) level at which certain material properties, such as the Young's modulus E, are observed. Examples for materials commonly used in mechanical applications are sheet metal, polycrystalline metals (bulk material) or particle- and fiber-reinforced composites. While these materials appear homogeneous from a structural point of view, observations at other length scales show a different picture.

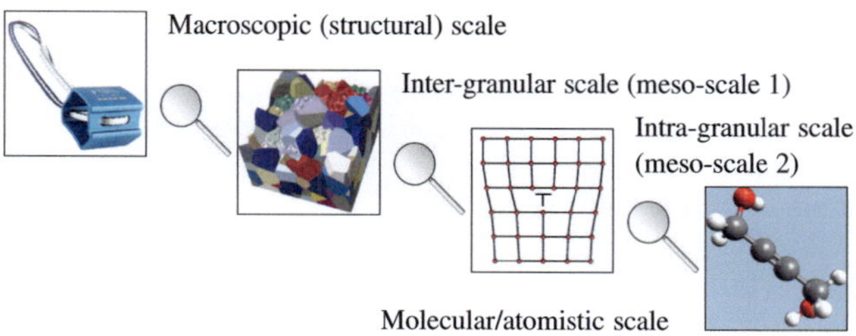

Figure 4.1: Schematic representation of the different length scales of a structure *

For example, a metallic part looks homogeneous at a length scale of meters (Fig. 4.1). Reducing the length scale to the sub-millimeter range, a granular structure is observed,

*Top left picture reprinted with permission of DMM International Ltd., Wales, UK, http://www.dmmclimbing.com.

where different grains can often be considered (almost) homogeneous in terms of their crystallographic orientation. When moving to the sub-micrometer level, individual dislocations and small precipitates can be observed. On an even finer (sub-)nanometer level individual atoms can be found.

The aim of this work is to study the thermo-mechanical response of such microheterogeneous materials. The latter is a topic with a rich history, see, e.g., the pioneering work of Voigt (1910) in which ideas relating the microscopic properties of single crystals and the constitutive response of a macroscopic aggregate are formulated. The consideration of all length scales is, usually, not possible in a unified framework. Nevertheless, recent advances in this direction have been made by Nikolov et al. (2010) for the linear-elastic properties of lobster cuticle. Due to the complex nature of these nested scale approaches their applicability to real materials is limited and an account for nonlinear material behavior is hardly possible. Therefore attention is confined to materials that allow for a two-scale description.

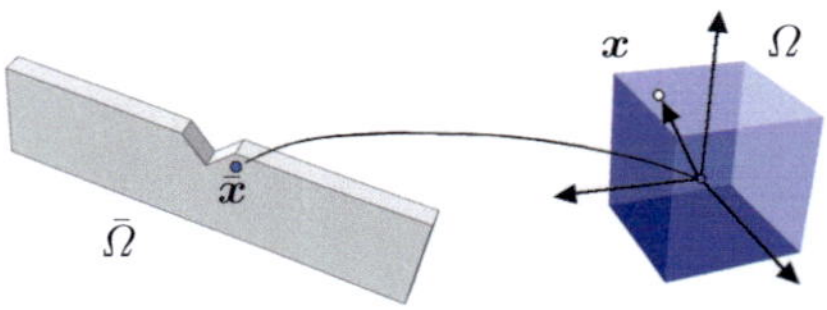

Figure 4.2: Local coordinate $\boldsymbol{x}$ and global coordinate $\bar{\boldsymbol{x}}$

To clarify the notation, the macroscopic domain is denoted by $\bar{\Omega}$ with points $\bar{\boldsymbol{x}}$. Each point $\bar{\boldsymbol{x}}$ is associated with a micro-volume $\Omega_{\text{tot}}(\bar{\boldsymbol{x}})$. The latter is decomposed into the subdomain Ω occupied by the material and the domain $\mathcal{H}$ occupied by the voids. It is worth emphasizing that the micro-volume may depend on $\bar{\boldsymbol{x}}$, i.e., the microstructure is not necessarily the same at each point. The center of each of the micro-volumes coincides with the macroscopic coordinate $\bar{\boldsymbol{x}}$, and the relative coordinate $\boldsymbol{x}$ is taken with respect to this point (Fig. 4.2).

Accounting for the voids the total material volume fraction of the microheterogeneous material is $c = |\Omega|/|\Omega_{\text{tot}}|$. The surface $\mathscr{S} = \partial\mathcal{H}$ denotes the ensemble of all internal surfaces due to voids or cracks. For convenience the normal vector $\boldsymbol{n}$ is assumed to point out of the pores on $\partial\mathcal{H}$. The entire boundary of the material is denoted $\Gamma = \partial\Omega$. Each of the N_{v} different voids with volume fraction $c_{\text{v},i}$ is assumed to be subjected to the hydrostatic pressure $p_{\text{v},i}$. Generally, quantities defined on the small scale are written without special typesetting, e.g., a field $\boldsymbol{w}$. If a counterpart on the macroscopic level exists, then it is over-lined, e.g., $\bar{\boldsymbol{w}}$. The volume averaging operator $\langle\bullet\rangle$ is extensively used in the following. Under the ergodicity hypothesis the volume average equals the ensemble average. It is defined for arbitrary additive quantities, like the internal and the free energy, the entropy, the dissipation, or the mass, via (Suquet, 1985a,b)

$$\bar{\bullet} = \langle\bullet\rangle = \langle\bullet\rangle_{\Omega_{\text{tot}}} = \frac{1}{|\Omega_{\text{tot}}|}\int_{\Omega_{\text{tot}}} \bullet \,\mathrm{d}V. \tag{4.1}$$

Note that not all quantities have a counterpart on the other scale, e.g., the plastic strain or hardening variables. As a counterexample one can consider a microstructure, where each micro-material has an individual scalar hardening variable associated with an individual physical effect. The effective material can *a priori* not be described by a macroscopic variable accounting for these states. However, if a macroscopic equivalent variable to a microscopic quantity exists, then the previous formula has to be modified for microstructures including voids. Additive quantities having a *zero-continuation* inside the voids can be averaged according to

$$\bar{\bullet} = \langle \bullet \rangle = \frac{1}{|\Omega_{\text{tot}}|} \int\limits_{\Omega} \bullet \, \mathrm{d}V = \frac{|\Omega|}{|\Omega_{\text{tot}}|} \frac{1}{|\Omega|} \int\limits_{\Omega} \bullet \, \mathrm{d}V = c \langle \bullet \rangle_{\Omega}. \tag{4.2}$$

This is true for all additive quantities such as the free energy, the entropy and the dissipation, but not for the stress or the strain. When considering a volume Ω_{tot} with boundary $\partial\Omega_{\text{tot}}$ from a macroscopic perspective, the only interaction of the domain with its surrounding is performed via its external surfaces. From this coarse scale perspective the stress tensor $\bar{\boldsymbol{\sigma}}$ and the displacement gradient $\bar{\boldsymbol{H}}$ are constant within the microscopic volume. For the macroscopic stress tensor Cauchy's lemma leads to

$$\bar{\boldsymbol{\sigma}} = \frac{1}{|\Omega_{\text{tot}}|} \int\limits_{\partial\Omega_{\text{tot}}} \boldsymbol{t} \otimes \boldsymbol{x} \, \mathrm{d}A. \tag{4.3}$$

Based on the Gauss theorem this surface integral can alternatively be expressed via

$$\bar{\boldsymbol{\sigma}} = \langle \boldsymbol{\sigma} \rangle = c \langle \boldsymbol{\sigma} \rangle_{\Omega} - \sum_{i=1}^{N_{\text{v}}} c_{\text{v},i} p_{\text{v},i} \boldsymbol{I} = c \langle \boldsymbol{\sigma} \rangle_{\Omega} + \bar{\boldsymbol{\sigma}}_{\text{c}}, \tag{4.4}$$

Then $\bar{\boldsymbol{\sigma}}_{\text{c}}$ is the total contribution of the applied pore pressure to the macroscopic stress. For the effective displacement gradient $\bar{\boldsymbol{H}}$ the displacement on the boundary of the microscopic volume is related to the macroscopic displacement gradient via

$$\bar{\boldsymbol{H}} = \frac{1}{|\Omega_{\text{tot}}|} \int\limits_{\partial\Omega_{\text{tot}}} \boldsymbol{u} \otimes \boldsymbol{n} \, \mathrm{d}A, \tag{4.5}$$

where $\boldsymbol{n}$ denotes the outer unit normal vector. Application of the Gauss theorem leads to

$$\bar{\boldsymbol{H}} = \frac{1}{|\Omega_{\text{tot}}|} \left(\int\limits_{\Omega} \operatorname{grad}(\boldsymbol{u}) \, \mathrm{d}V + \int\limits_{\mathscr{S}} \boldsymbol{u} \otimes \boldsymbol{n} \, \mathrm{d}A \right) = c \langle \boldsymbol{H} \rangle_{\Omega} + \bar{\boldsymbol{H}}_{\text{c}}. \tag{4.6}$$

Equation (4.6) also defines the effective strain tensor $\bar{\boldsymbol{\varepsilon}}$ via

$$\bar{\boldsymbol{\varepsilon}} = \operatorname{sym}(\bar{\boldsymbol{H}}) = c \langle \boldsymbol{\varepsilon} \rangle_{\Omega} + \operatorname{sym}(\bar{\boldsymbol{H}}_{\text{c}}) = c \langle \boldsymbol{\varepsilon} \rangle_{\Omega} + \bar{\boldsymbol{\varepsilon}}_{\text{c}}. \tag{4.7}$$

The part $\bar{\boldsymbol{\varepsilon}}_{\text{c}}$ associated with the inner surfaces is also referred to as cavity strain. In order to formally describe the considered class of quasi-static two-scale problems, the following hypotheses are assumed for the rest of this work:

[H1] A geometrically linear description of the material is used on both scales.

[H2] The typical length scale L of the macroscopic scale is significantly larger than the characteristic microscopic dimension l, i.e. $L \gg l$.

[H3] In order to allow for a continuum description of the small scale problem, it is required that the microscopic length scale l is clearly separated from the nano-scale at which molecular dynamics or alike become necessary.

[H4] The displacement $\bar{\boldsymbol{u}}$ and temperature $\bar{\theta}$ on the macroscopic scale have a counterpart on the microscale. It is defined based on the displacement $\bar{\boldsymbol{u}}(\bar{\boldsymbol{x}})$ and temperature $\bar{\theta}(\bar{\boldsymbol{x}})$ of the macroscopic point $\bar{\boldsymbol{x}}$ defining the barycenter of Ω_{tot} according to

$$\boldsymbol{u}(\boldsymbol{x}) = \bar{\boldsymbol{u}}(\bar{\boldsymbol{x}}) + \text{grad}_{\bar{\boldsymbol{x}}}(\bar{\boldsymbol{u}})\boldsymbol{x} + \tilde{\boldsymbol{u}}(\boldsymbol{x}), \qquad \langle \tilde{\boldsymbol{u}} \otimes \boldsymbol{n} \rangle_{\mathscr{S}} = \boldsymbol{0}, \tag{4.8}$$

$$\theta(\boldsymbol{x}) = \bar{\theta}(\bar{\boldsymbol{x}}) + \text{grad}_{\bar{\boldsymbol{x}}}(\bar{\theta})\boldsymbol{x} + \tilde{\theta}(\boldsymbol{x}), \qquad \langle \tilde{\theta} \boldsymbol{n} \rangle_{\mathscr{S}} = \boldsymbol{0}. \tag{4.9}$$

The constraints imposed on the fluctuations $\tilde{\boldsymbol{u}}, \tilde{\theta}$ guarantee that the macroscopic strain and the temperature gradient are replicated upon averaging.

[H5] The time scale of the macroscopic problem is clearly separated from the microscopic one. In particular, the rate of change of the macroscopic fields is assumed to be slow enough such that the associated micro-field can be treated as stationary.

[H6] The macroscopic problem is assumed quasi-static, i.e. the second order time derivative of the displacement fields on both scales satisfies for any norm $\| \bullet \|_A$

$$\left\| \frac{\partial^2 \bar{\boldsymbol{u}}}{\partial t \partial t} \right\|_A \approx 0 \approx \left\| \frac{\partial^2 \boldsymbol{u}}{\partial t \partial t} \right\|_A . \tag{4.10}$$

[H7] All micro-constituents can be described in the framework of generalized standard materials (GSM, see section 2.2.3). The rate of internal variables is

$$\dot{\boldsymbol{\xi}} = \begin{cases} \dfrac{\partial \phi^*}{\partial \boldsymbol{\mathcal{F}}} & \text{(rate-dependent)}, \\ \dot{\lambda} \dfrac{\partial \varphi}{\partial \boldsymbol{\mathcal{F}}} & \text{(rate-independent)}. \end{cases} \tag{4.11}$$

[H8] The specific force density $\boldsymbol{b}$ and the heat supply h are zero on the microscopic scale, i.e., zero source terms are assumed.

Note that due to **[H1]** the product of a gradient with respect to $\bar{\boldsymbol{x}}$ multiplied by the microscopic position vector $\boldsymbol{x}$ tends to zero. With **[H4]** the value of the microscopic temperature and displacement field can be approximated by the respective macroscopic value

$$\theta(\boldsymbol{x}) \approx \bar{\theta}(\bar{\boldsymbol{x}}), \qquad \boldsymbol{u}(\boldsymbol{x}) \approx \bar{\boldsymbol{u}}(\bar{\boldsymbol{x}}) \quad \forall \boldsymbol{x} \in \Omega(\bar{\boldsymbol{x}}). \tag{4.12}$$

The previous result can rigorously be obtained using asymptotic expansion as $l/L \to 0$ for periodic materials (see, e.g., section 5.2.4 in Besson et al., 2010). Given these assumptions and making use of (4.12), the balance of linear momentum and energy defining the partial differential equations for the microscale are (see also (2.49), (2.29) and section 2.2.3)

$$(P)\begin{cases} \operatorname{div}_{\boldsymbol{x}}(\boldsymbol{\sigma}) &= \mathbf{0} \qquad \text{in } \Omega \\ -\theta\dfrac{\partial\boldsymbol{\sigma}}{\partial\theta}\cdot\dot{\boldsymbol{\varepsilon}} + \left(\boldsymbol{\mathcal{F}} - \bar{\theta}\dfrac{\partial\boldsymbol{\mathcal{F}}}{\partial\theta}\right)\cdot\dot{\boldsymbol{\xi}} - \operatorname{div}_{\boldsymbol{x}}(\boldsymbol{q}) &= -\varrho\bar{\theta}\dfrac{\partial^2\psi}{\partial\theta\partial\theta}\dot{\bar{\theta}}, \quad \text{in } \Omega \\ + \text{ microscopic boundary conditions.} \end{cases} \tag{4.13}$$

Replacing the respective variables by their effective values and taking the derivatives with respect to the macroscopic scale, the set of differential equations on the structural level reads

$$(\bar{P})\begin{cases} \operatorname{div}_{\bar{\boldsymbol{x}}}(\bar{\boldsymbol{\sigma}}) &= \mathbf{0} \qquad \text{in } \bar{\Omega} \\ -\bar{\theta}\langle\dfrac{\partial\boldsymbol{\sigma}}{\partial\theta}\cdot\dot{\boldsymbol{\varepsilon}}\rangle + \left\langle\left(\boldsymbol{\mathcal{F}} - \bar{\theta}\dfrac{\partial\boldsymbol{\mathcal{F}}}{\partial\theta}\right)\cdot\dot{\boldsymbol{\xi}}\right\rangle & \\ -\operatorname{div}_{\bar{\boldsymbol{x}}}(\bar{\boldsymbol{q}}) + \bar{h} &= -\bar{\theta}\langle\varrho\dfrac{\partial^2\psi}{\partial\theta\partial\theta}\rangle\dot{\bar{\theta}} \quad \text{in } \bar{\Omega} \\ + \text{ macroscopic boundary conditions.} \end{cases} \tag{4.14}$$

The two sets of differential equations are coupled via the following quantities (Fig. 4.3), with the macroscopic mass density $\bar{\varrho} = \langle\varrho\rangle$

the local and effective stress (4.4)	$\boldsymbol{\sigma} \longleftrightarrow \bar{\boldsymbol{\sigma}}$
the local and effective strain (4.7)	$\boldsymbol{\varepsilon} \longleftrightarrow \bar{\boldsymbol{\varepsilon}}$
the local and effective temperature gradient	$\boldsymbol{g} \longleftrightarrow \bar{\boldsymbol{g}}$
the effective heat conductivity	$\boldsymbol{q} \longleftrightarrow \bar{\boldsymbol{q}}$
the thermo-elastic entropy rate	$\bar{\varrho}\dot{\bar{\eta}}^{\mathrm{e}} = -\bar{\theta}\langle\dfrac{\partial\boldsymbol{\sigma}}{\partial\theta}\cdot\dot{\boldsymbol{\varepsilon}}\rangle$
the mechanical dissipation	$\mathcal{D}_{\mathrm{p}} = \langle\boldsymbol{\mathcal{F}}\cdot\dot{\boldsymbol{\xi}}\rangle$
the thermo-plastic entropy rate	$\bar{\varrho}\dot{\bar{\eta}}^{\mathrm{p}} = -\bar{\theta}\langle\dfrac{\partial\boldsymbol{\mathcal{F}}}{\partial\theta}\cdot\dot{\boldsymbol{\xi}}\rangle$
the effective heat capacity	$\bar{\varrho}\bar{c}_{\theta} = -\bar{\theta}\langle\varrho\dfrac{\partial^2\psi}{\partial\theta\partial\theta}\rangle$

Note that the microscopic boundary conditions relate to the first two points in this list. A detailled discussion is presented in the following section.

4.1.1 Boundary conditions for homogenization methods

While the macroscopic boundary conditions are addressed in section 2.2.1, the microscopic boundary conditions have not been specified, yet. In fact, they constitute an important

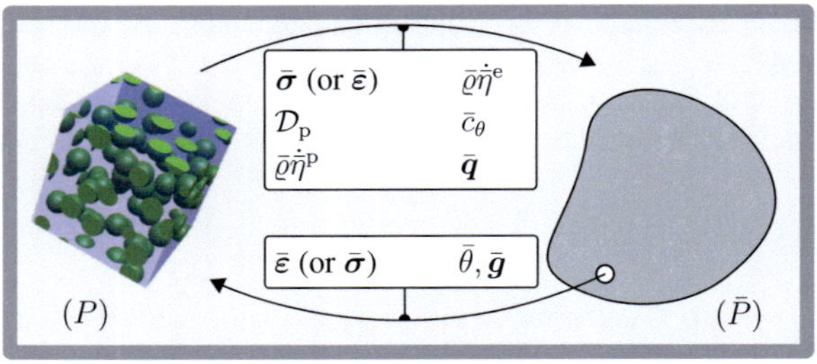

Figure 4.3: Quantities involved in the thermo-mechanically coupled scale transition

part of the scale coupling that relates (P) and $(\bar{P})$. The examination of the influence of the type of boundary conditions on the computational results has a long history and many contributions have been published with the aim of examining their significance (see, e.g., Ostoja-Starzewski, 2001; Miehe, 2002; Fritzen and Böhlke, 2010a). A natural requirement imposed on microscopic boundary conditions in the context of homogenization is the equivalence of the macroscopic power and the volume average of its microscopic counterpart:

$$\langle \boldsymbol{\sigma} \cdot \dot{\varepsilon} \rangle = \langle \boldsymbol{\sigma} \rangle \cdot \langle \dot{\varepsilon} \rangle = \bar{\boldsymbol{\sigma}} \cdot \dot{\bar{\varepsilon}}. \tag{4.15}$$

The latter condition is well-known as the Hill-Mandel condition (Hill, 1963) and has been subject of many investigations, for instance in the work by Suquet (1985a). A result of (4.15) is the requirement of energetic orthogonality of the admissible displacement fluctuations $\tilde{\boldsymbol{u}}_*$ and admissible fluctuations in the tractions $\tilde{\boldsymbol{t}}_*$ according to

$$\int_\Gamma \dot{\tilde{\boldsymbol{u}}}_* \cdot \tilde{\boldsymbol{t}}_* \, \mathrm{d}A = 0. \tag{4.16}$$

It was shown by Suquet (1985a) that the following boundary conditions satisfy (4.15):

- Uniform kinematic boundary conditions (ukbc)

$$\boldsymbol{u} = \bar{\boldsymbol{H}}\boldsymbol{x} \quad \text{on } \partial\Omega_{\rm tot} \cap \partial\Omega, \tag{4.17}$$

- Uniform static boundary conditions (usbc)

$$\boldsymbol{t} = \bar{\boldsymbol{\sigma}}\boldsymbol{n} \quad \text{on } \partial\Omega_{\rm tot} \cap \partial\Omega, \tag{4.18}$$

- Periodic fluctuation conditions

$$\boldsymbol{u} = \bar{\boldsymbol{H}}\boldsymbol{x} + \tilde{\boldsymbol{u}}, \qquad \tilde{\boldsymbol{u}}(\boldsymbol{x}_+) = \tilde{\boldsymbol{u}}(\boldsymbol{x}_-), \tag{4.19}$$

$$\boldsymbol{t} = \bar{\boldsymbol{\sigma}}\boldsymbol{n} + \tilde{\boldsymbol{t}}, \qquad \tilde{\boldsymbol{t}}(\boldsymbol{x}_+) = -\tilde{\boldsymbol{t}}(\boldsymbol{x}_-), \tag{4.20}$$

with $\boldsymbol{x}_+, \boldsymbol{x}_-$ being elements of the periodic point sets Γ_+ and Γ_- on the boundary. Either the macroscopic strain $\bar{\varepsilon}$, the macroscopic stress $\bar{\boldsymbol{\sigma}}$ or a combination of both can be prescribed. Accordingly it is referred to periodic kinematic, static and mixed boundary conditions, respectively (pkbc, psbc, pmbc). A two-dimensional example of periodic displacement field and its decomposition into the homogeneous deformation and the fluctuation part are shown in Fig. 4.4.

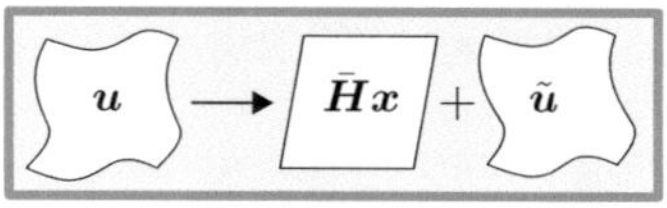

Figure 4.4: Two-dimensional example of a periodic displacement fluctuation field

Thermal boundary conditions
In the context of thermal homogenization, the classification of the boundary conditions follows a similar line as for mechanical boundary conditions. Here the solution variable is θ instead of $\boldsymbol{u}$ and the quantity of interest is the heat flux $\boldsymbol{q}$, which is related to the gradient $\boldsymbol{g}$ of the temperature θ. Periodic temperature fluctuations are chosen throughout this work, i.e., for a given macroscopic temperature gradient $\bar{\boldsymbol{g}}$, the fluctuations $\tilde{\theta}$ of the temperature satisfy

$$\tilde{\theta}(\boldsymbol{x}_+) - \tilde{\theta}(\boldsymbol{x}_-) = 0 \quad \forall \boldsymbol{x}_\pm \in \Gamma_\pm. \tag{4.21}$$

The absolute value $\bar{\theta}$ of the mean temperature on the microscale can numerically be enforced based on the content of stored thermal energy.

4.2 Linear homogenization problems

When considering the specific case of isothermal, elastic deformation, then the balance of linear momentum in (4.13) is a linear partial differential equation. The same holds for the balance of energy considering elastic, isothermal heat conduction at constant deformation. In general, if a d-dimensional set of independent boundary conditions $\mathcal{B}^{(\alpha)}$ is considered in a linear partial differential equation and the solutions to the individual problems are denoted $\boldsymbol{u}^{(\alpha)}$, then for any linear combination of the boundary conditions

$$\mathcal{B}_* = \sum_{\alpha=1}^{d} k_\alpha \mathcal{B}^{(\alpha)}, \qquad k_\alpha \in \mathbb{R}, \tag{4.22}$$

the solution $\boldsymbol{u}_*(\hat{k}, \boldsymbol{x})$ is readily obtained by superposition

$$\boldsymbol{u}_*(\hat{k}, \boldsymbol{x}) = \sum_{\alpha=1}^{d} k_\alpha \boldsymbol{u}^{(\alpha)}. \tag{4.23}$$

Due to the linearity of the averaging operator $\langle \bullet \rangle$, this holds true for all averages of linear functions of $\boldsymbol{u}_*$, i.e. for the gradient and linear functions of the gradient.

Linear elastic homogenization. In the case of linear elastic homogenization, the effective stress-strain relation

$$\bar{\boldsymbol{\sigma}} = \bar{\mathbb{C}}[\bar{\varepsilon}] \tag{4.24}$$

is sought-after. Since both, $\bar{\boldsymbol{\sigma}}$ and $\bar{\varepsilon}$ are symmetric second order tensors, a total of six different boundary conditions has to be considered to determine the effective fourth order stiffness tensor $\bar{\mathbb{C}}$. So far, the requirement imposed on the six boundary conditions is their linear independence. After solution of the respective isothermal microscopic problems $P^{(\alpha)}$, one can associate with each of the solutions an effective stress $\bar{\boldsymbol{\sigma}}^{(\alpha)}$ and an effective strain $\bar{\varepsilon}^{(\alpha)}$. Using the six-dimensional vector-matrix representation for second and fourth order tensors introduced in appendix A with the orthonormal basis $\{\boldsymbol{B}^{(\alpha)}\}_{\alpha=1,\ldots,6}$, a linear equation of the type

$$(\hat{E})_{\alpha\beta} = \boldsymbol{B}^{(\alpha)} \cdot \bar{\varepsilon}^{(\beta)}, \qquad (\hat{\Sigma})_{\alpha\beta} = \boldsymbol{B}^{(\alpha)} \cdot \bar{\boldsymbol{\sigma}}^{(\beta)}, \tag{4.25}$$

$$(\hat{\hat{C}})_{\alpha\beta} = \bar{\mathbb{C}} \cdot \boldsymbol{B}^{(\alpha)} \otimes \boldsymbol{B}^{(\beta)}, \qquad \hat{\hat{C}}\hat{E} = \hat{\Sigma}, \tag{4.26}$$

is obtained. The effective stiffness tensor $\bar{\mathbb{C}}$ of the material is obtained in terms of its matrix representation $\hat{\hat{C}}$ from

$$\hat{\hat{C}} = \hat{\Sigma}\hat{E}^{-1} \qquad \bar{\mathbb{C}} = \sum_{\alpha=1}^{6}\sum_{\beta=1}^{6}(\hat{\hat{C}})_{\alpha\beta}\boldsymbol{B}^{(\alpha)} \otimes \boldsymbol{B}^{(\beta)}. \tag{4.27}$$

An arbitrary deformation $\bar{\varepsilon} = \sum_{\alpha=1}^{6} \bar{\varepsilon}_\alpha \boldsymbol{B}^{(\alpha)}$ imposed on the unit volume then leads to the effective stress response $\bar{\mathbb{C}}[\bar{\varepsilon}]$. Moreover, the local strain field and the local stress field are recovered by

$$\boldsymbol{\varepsilon}(\boldsymbol{x}, \hat{\bar{\varepsilon}}) = \sum_{\alpha=1}^{N}(\hat{E}^{-1})_{\alpha\beta}\bar{\varepsilon}_\beta \boldsymbol{\varepsilon}^{(\alpha)}(\boldsymbol{x}), \qquad \boldsymbol{\sigma}(\boldsymbol{x}, \hat{\bar{\varepsilon}}) = \sum_{\alpha=1}^{N}(\hat{E}^{-1})_{\alpha\beta}\bar{\varepsilon}_\beta \boldsymbol{\sigma}^{(\alpha)}(\boldsymbol{x}). \tag{4.28}$$

Equivalently, the introduction of a prescribed macroscopic stress $\bar{\boldsymbol{\sigma}} = \sum_{\alpha=1}^{6} \bar{\sigma}_\alpha \boldsymbol{B}^{(\alpha)}$ leads to

$$\boldsymbol{\varepsilon}(\boldsymbol{x}, \hat{\bar{\sigma}}) = \sum_{\alpha=1}^{N}(\hat{\Sigma}^{-1})_{\alpha\beta}\bar{\sigma}_\beta \boldsymbol{\varepsilon}^{(\alpha)}(\boldsymbol{x}), \qquad \boldsymbol{\sigma}(\boldsymbol{x}, \hat{\bar{\sigma}}) = \sum_{\alpha=1}^{N}(\hat{\Sigma}^{-1})_{\alpha\beta}\bar{\sigma}_\beta \boldsymbol{\sigma}^{(\alpha)}(\boldsymbol{x}). \tag{4.29}$$

Based on (4.28) and (4.29) the strain and the stress localization operators

$$\mathbb{A}(\boldsymbol{x}) = \sum_{\alpha=1}^{N} \boldsymbol{\varepsilon}^{(\alpha)}(\boldsymbol{x})(\hat{E}^{-1})_{\alpha\beta} \otimes \boldsymbol{B}^{(\beta)}, \qquad \boldsymbol{\varepsilon}(\boldsymbol{x}, \bar{\varepsilon}) = \mathbb{A}(\boldsymbol{x})[\bar{\varepsilon}], \tag{4.30}$$

$$\mathbb{B}(\boldsymbol{x}) = \sum_{\alpha=1}^{N} \boldsymbol{\sigma}^{(\alpha)}(\boldsymbol{x})(\hat{\Sigma}^{-1})_{\alpha\beta} \otimes \boldsymbol{B}^{(\beta)}, \qquad \boldsymbol{\sigma}(\boldsymbol{x}, \bar{\boldsymbol{\sigma}}) = \mathbb{B}(\boldsymbol{x})[\bar{\boldsymbol{\sigma}}] \tag{4.31}$$

are defined. Note, that based on prescribed orthonormal stresses or strains the matrices $\hat{\Sigma}$ and $\hat{E}$ are identity matrices and the expressions (4.30), (4.31) simplify considerably. In the case of prescribed orthonormal strains $\bar{\boldsymbol{\varepsilon}}^{(\alpha)} = \boldsymbol{B}^{(\alpha)}$ the strain localization operator (4.28) becomes

$$\mathbb{A}(\boldsymbol{x}) = \sum_{\alpha=1}^{6} \boldsymbol{\varepsilon}^{(\alpha)}(\boldsymbol{x}) \otimes \boldsymbol{B}^{(\alpha)}. \tag{4.32}$$

Analogously, prescribed orthogonal stresses $\bar{\boldsymbol{\sigma}}^{(\alpha)} = \sigma_0 \boldsymbol{B}^{(\alpha)}$ in (4.29) lead to

$$\mathbb{B}(\boldsymbol{x}) = \frac{1}{\sigma_0} \sum_{\alpha=1}^{6} \boldsymbol{\sigma}^{(\alpha)}(\boldsymbol{x}) \otimes \boldsymbol{B}^{(\alpha)}. \tag{4.33}$$

For completeness the displacement localization operator under prescribed strains is

$$\boldsymbol{u}(\boldsymbol{x}, \bar{\boldsymbol{\varepsilon}}) = \mathbb{Y}(\boldsymbol{x})[\bar{\boldsymbol{\varepsilon}}], \qquad \mathbb{Y}(\boldsymbol{x}) = \sum_{\alpha=1}^{6} \boldsymbol{u}^{(\alpha)}(\boldsymbol{x})(\hat{E}^{-1})_{\alpha\beta} \otimes \boldsymbol{B}^{(\beta)}. \tag{4.34}$$

Linear thermal homogenization. The procedure for the determination of the effective thermal conductivity follows the same line as described for the linear elastic properties. For brevity the case of boundary conditions imposing three orthogonal temperature gradients $\bar{\boldsymbol{g}}^{(\alpha)}$ of amplitude g_0 are assumed. The fields solving the respective problem are denoted $\theta^{(\alpha)}(\boldsymbol{x}), \boldsymbol{g}^{(\alpha)}(\boldsymbol{x}), \boldsymbol{q}^{(\alpha)}(\boldsymbol{x})$. Then the localization rule for the temperature gradient reads

$$\boldsymbol{g}(\boldsymbol{x}, \bar{\boldsymbol{g}}) = \left(\sum_{\alpha=1}^{3} \frac{1}{g_0^2} \boldsymbol{g}^{(\alpha)}(\boldsymbol{x}) \otimes \bar{\boldsymbol{g}}^{(\alpha)} \right) \bar{\boldsymbol{g}}. \tag{4.35}$$

Assuming Fourier type heat conduction as in (2.48) with the conductivity tensor $\boldsymbol{\kappa}$ gives

$$\bar{\boldsymbol{\kappa}} = \sum_{\alpha=1}^{3} \frac{1}{g_0^2} \langle \boldsymbol{\kappa} \boldsymbol{g}^{(\alpha)}(\boldsymbol{x}) \rangle \otimes \bar{\boldsymbol{g}}^{(\alpha)}, \qquad \bar{\boldsymbol{q}}(\bar{\boldsymbol{g}}) = -\bar{\boldsymbol{\kappa}} \bar{\boldsymbol{g}}. \tag{4.36}$$

Linear thermo-elasticity. The linear thermo-elastic properties are determined based on an elastic eigenstress problem. Here the case of zero prescribed strain and the thermal eigenstress field

$$\boldsymbol{\tau}_\theta(\boldsymbol{x}) = \boldsymbol{\beta}_\theta(\boldsymbol{x}) \Delta\theta_0, \tag{4.37}$$

with the temperature increment $\Delta\theta_0 \neq 0$ are considered. The thereby induced linear-elastic eigenstress problem reads

$$\operatorname{div}(\mathbb{C}[\boldsymbol{\varepsilon}_\theta] + \boldsymbol{\tau}_\theta]) = \boldsymbol{0}, \qquad \langle \boldsymbol{\varepsilon}_\theta \rangle = \boldsymbol{0}. \tag{4.38}$$

An alternative description with static boundary conditions is possible, but is omitted for brevity. The local stress field solving (4.38) is denoted by

$$\boldsymbol{\sigma}_\theta(\boldsymbol{x}) = \mathbb{C}(\boldsymbol{x})[\boldsymbol{\varepsilon}_\theta(\boldsymbol{x})] + \boldsymbol{\beta}_\theta(\boldsymbol{x})\Delta\theta_0 \tag{4.39}$$

and the corresponding displacement field is referred to as $\boldsymbol{u}_\theta$. The resulting effective stress $\bar{\boldsymbol{\sigma}}_\theta = \langle \boldsymbol{\sigma}_\theta \rangle$ determines the effective thermal eigenstress $\bar{\boldsymbol{\beta}}_\theta$ and the effective thermal expansion coefficient $\bar{\boldsymbol{\alpha}}_\theta$ of the micro-structured material via

$$\bar{\boldsymbol{\beta}}_\theta = \frac{1}{\Delta\theta_0}\bar{\boldsymbol{\sigma}}_\theta, \qquad \bar{\boldsymbol{\alpha}}_\theta = -\bar{\mathbb{C}}^{-1}[\bar{\boldsymbol{\beta}}_\theta]. \tag{4.40}$$

The localization rule for the displacement, the strain and the stress tensor with respect to a prescribed temperature increment $\Delta\bar{\theta} = \bar{\theta} - \bar{\theta}_0$ from the reference temperature $\bar{\theta}_0$ and a prescribed macroscopic strain $\bar{\varepsilon}$ are

$$\boldsymbol{u}(\boldsymbol{x}, \bar{\varepsilon}, \Delta\bar{\theta}) = \mathbb{Y}(\boldsymbol{x})[\bar{\varepsilon}] + \boldsymbol{u}_\theta(\boldsymbol{x})\frac{\Delta\bar{\theta}}{\Delta\theta_0}, \tag{4.41}$$

$$\boldsymbol{\varepsilon}(\boldsymbol{x}, \bar{\varepsilon}, \Delta\bar{\theta}) = \mathbb{A}(\boldsymbol{x})[\bar{\varepsilon}] + \boldsymbol{\varepsilon}_\theta(\boldsymbol{x})\frac{\Delta\bar{\theta}}{\Delta\theta_0} \tag{4.42}$$

$$\boldsymbol{\sigma}(\boldsymbol{x}, \bar{\varepsilon}, \Delta\bar{\theta}) = \mathbb{C}(\boldsymbol{x})\mathbb{A}(\boldsymbol{x})[\bar{\varepsilon}] + \boldsymbol{\sigma}_\theta(\boldsymbol{x})\frac{\Delta\bar{\theta}}{\Delta\theta_0}. \tag{4.43}$$

The effective constitutive response accounting for (4.27), (4.36) and (4.40) is given by

$$\bar{\boldsymbol{\sigma}}(\bar{\varepsilon}, \Delta\bar{\theta}) = \bar{\mathbb{C}}[\bar{\varepsilon}] + \bar{\boldsymbol{\beta}}_\theta\Delta\bar{\theta}, \qquad \bar{\boldsymbol{q}}(\bar{\boldsymbol{g}}) = -\bar{\boldsymbol{\kappa}}\bar{\boldsymbol{g}}. \tag{4.44}$$

For completeness, the effective heat capacity is also provided

$$\bar{c}_\theta = \frac{\bar{\theta}}{\bar{\varrho}}\Big\langle -\varrho\frac{\partial^2\psi}{\partial\theta\partial\theta}\Big\rangle. \tag{4.45}$$

While these quantities are usually used in engineering applications, the actual homogenized thermo-elastic response is more complicated due to the thermo-elastic entropic effects mentioned in section 4.1. These are commonly referred to as Gough-Joule effect for bulk materials. The exact form of the homogenized balance of energy in a thermo-elastic setting reads

$$-\bar{\theta}\langle\frac{\partial\boldsymbol{\sigma}}{\partial\theta}\cdot\dot{\boldsymbol{\varepsilon}}\rangle - \mathrm{div}_{\bar{\boldsymbol{x}}}(\bar{\boldsymbol{q}}) + \bar{h} = \bar{\varrho}\bar{c}_\theta\dot{\bar{\theta}}.$$

Expanding the first term together with the rate form of (4.42) gives

$$\langle\frac{\partial\boldsymbol{\sigma}}{\partial\theta}\cdot\dot{\boldsymbol{\varepsilon}}\rangle = \frac{1}{\Delta\theta_0}\left(\langle\mathbb{A}^{\mathsf{T}}[\boldsymbol{\sigma}_\theta]\rangle\cdot\dot{\bar{\varepsilon}} + \langle\boldsymbol{\sigma}_\theta\cdot\boldsymbol{\varepsilon}_\theta\rangle\dot{\bar{\theta}}\right). \tag{4.46}$$

From the Hill-Mandel condition the relation

$$\langle \mathbb{A}^{\mathsf{T}}[\boldsymbol{\sigma}] \rangle = \langle \boldsymbol{\sigma} \rangle \tag{4.47}$$

can be derived for all statically admissible stress fields $\boldsymbol{\sigma}$, i.e. for $\operatorname{div}(\boldsymbol{\sigma}) = 0$ in Ω. Therefore, the representation (4.30) of the strain localization operator with $\langle \boldsymbol{\varepsilon}^{(\alpha)} \rangle = \boldsymbol{B}^{(\alpha)}$ is inserted into (4.47) to obtain

$$\langle \mathbb{A}^{\mathsf{T}}[\boldsymbol{\sigma}] \rangle = \sum_{\alpha=1}^{6} \boldsymbol{B}^{(\alpha)} \langle \boldsymbol{\varepsilon}^{(\alpha)} \cdot \boldsymbol{\sigma} \rangle = \sum_{\alpha=1}^{6} \boldsymbol{B}^{(\alpha)} \langle \boldsymbol{\varepsilon}^{(\alpha)} \rangle \cdot \langle \boldsymbol{\sigma} \rangle = \langle \boldsymbol{\sigma} \rangle. \tag{4.48}$$

Incorporating $\langle \boldsymbol{\varepsilon}_\theta \rangle = \mathbf{0}$ the identity

$$\left\langle \frac{\partial \boldsymbol{\sigma}}{\partial \theta} \cdot \dot{\boldsymbol{\varepsilon}} \right\rangle = \frac{1}{\Delta\theta_0} \langle \boldsymbol{\sigma}_\theta \rangle \cdot \dot{\bar{\boldsymbol{\varepsilon}}} \tag{4.49}$$

is attained and a simplified representation of (4.46) is

$$-\bar{\theta}\bar{\boldsymbol{\beta}}_\theta \cdot \dot{\bar{\boldsymbol{\varepsilon}}} - \operatorname{div}_{\bar{\boldsymbol{x}}}(\bar{\boldsymbol{q}}) + \bar{h} = \bar{\varrho}\bar{c}_\theta \dot{\bar{\theta}}. \tag{4.50}$$

4.3 Semi-analytical methods

In a general context, methods for the determination of the effective material response can be characterized following Fig. 4.5. Note that this characterization is valid not only for the linear case examined to this point, but also for the nonlinear regime. Three major groups can be identified: (i) (semi-)analytical methods, (ii) hybrid methods and (iii) computational methods. In the (semi-)analytical methods the localization operators $\mathbb{A}, \mathbb{B}$ of the real material are replaced by analytical representations. The methods can further be sub-divided into methods referring to purely statistical descriptions of the material and into methods referring to analytical solutions of reference problems. They have in common that the representation (4.27) of the effective stiffness operator is not necessarily used, but it is expressed in terms the phase-averages of $\mathbb{A}$ (or $\mathbb{B}$) and the phase-wise constant stiffness tensor. Similar constructions are possible for the thermal conductivity problem. Most notably, the actual solution of the microstructural problem is usually not computed, i.e. no local fields are determined. Based on appropriate assumptions, the real problem is replaced by (often a series of) simpler problems with the aim of providing an approximate solution that provides a sufficient accuracy. The term *semi* accounts for the fact that the solutions can often not be given in closed form, but for example in terms of implicit and usually nonlinear equations or in terms of integrals that need to be calculated numerically. An important sub-class of the semi-analytical methods are bounds on the strain energy of the material which result in limitations of the effective elastic material response. In the following, several popular semi-analytical methods are briefly described. Detailed monographs on the topic have been written, e.g., by Nemat-Nasser and Hori (1999); Torquato (2002); Qu and Cherkaoui (2006).

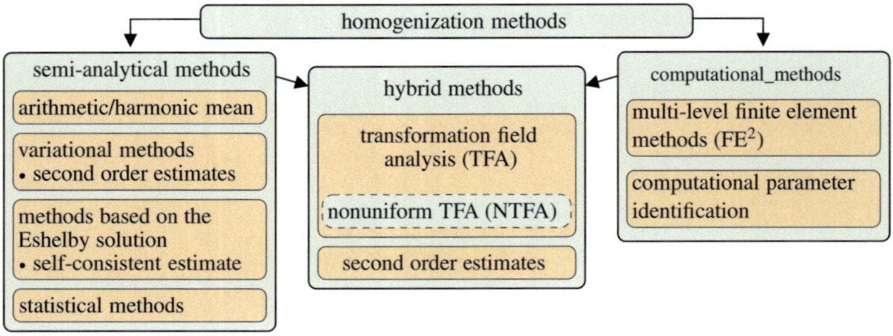

Figure 4.5: Schematic characterization of homogenization methods

An estimate of the elastic stiffness of a material was proposed by Voigt (1910) based on the assumption of a homogeneous deformation throughout the microstructured material. Asserting a homogeneous distribution of the applied strain, the effective stiffness tensor of such a material is given by the arithmetic mean of the phase stiffness $\mathbb{C}_\alpha$ weighted by the phase volume fraction c_α

$$\mathbb{C}_{\mathrm{V}} = \sum_{\alpha=1}^{N} c_\alpha \mathbb{C}_\alpha, \qquad \mathbb{S}_{\mathrm{V}} = \mathbb{C}_{\mathrm{V}}^{-1}. \tag{4.51}$$

Asserting a homogeneous stress distribution within the material, Reuss (1929) found that the arithmetic mean of the compliance tensors of the phases provides an estimate of the effective compliance tensor

$$\mathbb{S}_{\mathrm{R}} = \sum_{\alpha=1}^{N} c_\alpha \mathbb{S}_\alpha = \sum_{\alpha=1}^{N} c_\alpha \mathbb{C}_\alpha^{-1}, \qquad \mathbb{C}_{\mathrm{R}} = \mathbb{S}_{\mathrm{R}}^{-1}. \tag{4.52}$$

If the microstructure contains voids or porosity, the Reuss estimate does no longer apply. While the Voigt estimate has the advantage of being well-defined for all materials, i.e. in the presence of pores, the Reuss estimate fails for porous materials due to the non-existence of the compliance tensor of the phases. It was found by Hill (1952) that the two predictions actually constitute bounds for the possible effective material response by

$$\mathbb{C}_{\mathrm{R}} \leq \bar{\mathbb{C}} \leq \mathbb{C}_{\mathrm{V}}, \qquad \mathbb{S}_{\mathrm{R}} \geq \bar{\mathbb{S}} \geq \mathbb{S}_{\mathrm{V}}, \tag{4.53}$$

with the inequality understood in the spectral sense. While the simplicity of the arithmetic (or harmonic) mean are appealing, it is clear, e.g., by consideration of Fig. 3.1, that the micromorphology and topology of the material has an important influence on the effective material response. This influence cannot be considered by the Voigt and Reuss estimate. Additionally, the range of the elastic properties spanned by the Voigt and Reuss bounds is prohibitively wide if the contrast in the elastic properties of the micro-components gets large. Thus, they may not provide useful estimates. This issue is overcome by more sophisticated homogenization schemes such as the estimate introduced by Hashin and Shtrikman (1962a,b). Assuming a quasi-homogeneous and quasi-isotropic material, a variational problem is derived which results in an upper and lower bound of the isotropic material parameters of the microheterogeneous material. These bounds have the advantage of predicting a significantly smaller range of the effective elastic parameters when compared to the Voigt and Reuss estimate. Willis (1977) generalized the results to more general materials with local anisotropic physical behavior and anisotropic spheroidal micromorphology.

Referring to a purely statistical description of the microstructure and initiated by the work of Kröner (1958), the more refined approaches presented by Dederichs and Zeller (1973); Zeller and Dederichs (1973); Kröner (1977) and others were established. They provide bounds on the effective elastic properties. It has numerically been verified that the bounds are satisfied for polycrystalline aggregates by Fritzen et al. (2009), if the morphology of the microstructure is sufficiently isotropic. Most notably, the bounds of fifth and higher order provide a small range for the effective material response for small and moderate contrast in the elastic properties of the constituents.

Besides these methods, the analytical solution for the elastic field of an ellipsoidal inclusion embedded in an elastic matrix developed in the seminal work of Eshelby (1957) is the basis for many homogenization methods. Most notably, the solution exists for arbitrary aspect ratios giving rise to methods dealing with real-life composites that often exhibit strongly anisotropic morphologies, e.g., in terms of prolate/oblate particles or fiber composites.

One of these methods is the Mori-Tanaka approximation (Mori and Tanaka, 1973) in which the inclusion strain and the matrix strain are related based on the Eshelby solution. While the method can be applied to thermal and electrical conductivity problems, it is not possible to apply the method to composites containing more than two different phases. Investigations on the application of staggered homogenization procedures involving the Mori-Tanaka scheme in order to overcome this deficiency are discussed by, e.g., Li (1999); Pierard et al. (2004). Additionally, issues concerning possible non-symmetry of the resulting effective stiffness tensor have been addressed by, e.g.,Benveniste et al. (1991).

Another approach is found in the self-consistent estimates (Kröner, 1958; Hill, 1965; Willis, 1977) in which the constituents are successively embedded into a matrix material having the properties of the effective medium (see also Walpole, 1966b,a). Note that the latter are a priori not known. In general, the procedure results in an implicit equation that has to be solved iteratively.

A similar approach was pursued by Christensen and Lo (1979) by the double-inclusion scheme in which the inclusion phase is surrounded by a layer of the matrix material. This composite sphere (or ellipsoidal inclusion) is then embedded in the (unknown) composite material. Later, generalizations have been made by Hervé and Zaoui (1993, 1995) to allow for multiple inclusion layers.

While the presented body of literature is commonly focused on the purely mechanical aspect, the introduction of a thermal eigenstress or polarization field as in (4.38) is often possible. This holds particularly true if constituent-wise constant polarizations are considered. Then the thermo-mechanical interactions can be computed using similar approaches as presented earlier. It was for example shown by Rosen and Hashin (1970) that the effective thermal expansion of a two-phase material with isotropic micro-constituents is given by

$$\bar{\boldsymbol{\alpha}}_\theta = \left((1-c)\alpha_{\mathrm{m}} + c\alpha_{\mathrm{p}} + \frac{\alpha_{\mathrm{m}} - \alpha_{\mathrm{p}}}{K_{\mathrm{m}}^{-1} - K_{\mathrm{p}}^{-1}} \left(\frac{1}{\bar{K}} - \frac{1-c}{K_{\mathrm{m}}} - \frac{c}{K_{\mathrm{p}}} \right) \right) \boldsymbol{I}, \tag{4.54}$$

where $\alpha_{\mathrm{m}}, \alpha_{\mathrm{p}}, K_{\mathrm{m}}, K_{\mathrm{p}}$ denote the isotropic thermal expansion coefficient and the bulk moduli of the two phases, $\bar{K}$ is the homogenized bulk modulus and c the concentration of the phase indexed with p. It should be pointed out that such exact representations can be useful in order to validate numerical implementations.

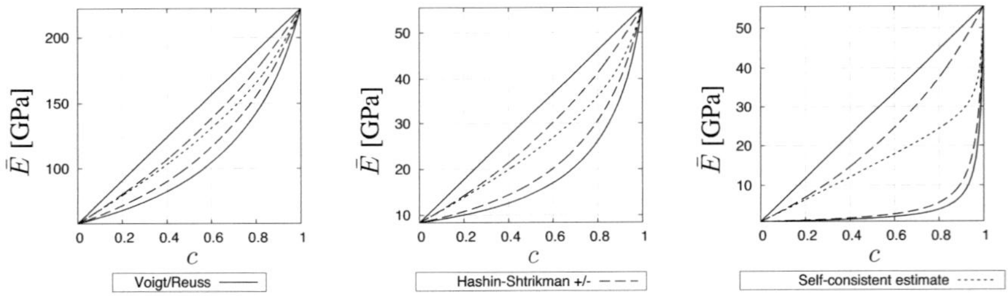

$E_0 = 70$ GPa, $E_1 = 400$ GPa $E_0 = 10$ GPa, $E_1 = 100$ GPa $E_0 = 1$ GPa, $E_1 = 100$ GPa

Figure 4.6: Comparison of the effective Young's modulus $\bar{E}$ for different semi-analytical estimates as function of the fraction c of the stiffer constituent and for different contrasts in the local Young's modulus

Examples of the Voigt and Reuss bound, the upper and lower Hashin-Shtrikman estimate and the self-consistent scheme are shown for the effective Young's modulus $\bar{E}$ and the derived Poisson ratio $\bar{\nu}$ in Fig. 4.6 and 4.7 for different ratios of the Young's modulus of the constituents E_0 and E_1. The values $\nu_0 = 0.3, \nu_1 = 0.2$ are identical for all graphs.

In the case of moderate phase contrast the Hashin-Shtrikman bounds provide rather close bounds on the elastic properties which are sufficient for many applications. As the phase contrast (here E_1/E_0) rises, the provided upper and lower bounds tend to the Voigt and Reuss estimate. The thereby obtained estimation of the real response of the material is not satisfactory for most applications. Note that the self-consistent estimate is always found within the upper and lower Hashin-Shtrikman prediction. If non-spheroidal microstructures are considered, e.g. for interconnected metal-ceramic composites, the assumptions entering the semi-analytical methods are often not valid. Consequently, exact predictions of the material behavior in these cases cannot be attained analytically.

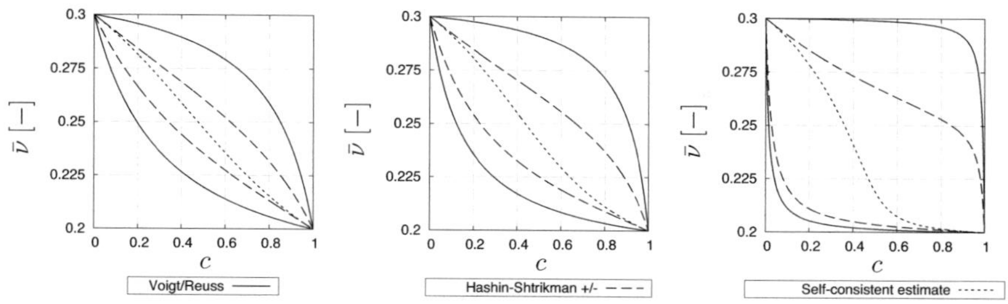

$E_0 = 70$ GPa, $E_1 = 400$ GPa $E_0 = 10$ GPa, $E_1 = 100$ GPa $E_0 = 1$ GPa, $E_1 = 100$ GPa

Figure 4.7: Comparison of the effective Poisson ratio $\bar{\nu}$ for different semi-analytical estimates as function of the fraction c of the stiffer constituent and for different contrasts in the local Young's modulus

Computational homogenization methods refer to a numerical solution of the underlying physical problem. Usually a discretization with respect to space and time is required before the actual computation can be performed. In the following section methods for linear problems are discussed and an account to purely computational approaches in the context of nonlinear material behavior is given in section 4.4.2.
The group of hybrid homogenization methods combines numerical solution methods with theoretical investigations. In particular the numerical methods provide data required for further utilization in a micro-mechanical context. This class of homogenization techniques is often applied in the context of physical and/or geometrical nonlinearity, where semi-analytical solutions are not sufficient and purely computational approaches require an inacceptable amount of resources. A discussion is found in section 4.4.2 and chapter 5 is completely devoted to the Nonuniform Transformation Field Analysis which belongs to this class.

4.4 Computational homogenization

4.4.1 Linear computational homogenization

In contrast to the semi-analytical methods computational methods refer to numerical solutions of the microscopic problem (P). Due to the linearity of the effective material response with respect to external loading, it suffices to compute seven quasi-static load cases and three stationary thermal conduction problems in order to resolve all localization operators exactly (see section 4.2). The load cases are defined by

[L1] six prescribed macroscopic strains $\bar{\boldsymbol{\varepsilon}}^{(\alpha)}$ $(\alpha = 1, \ldots, 6)$,

[L2] a thermal eigen-stress problem with $\boldsymbol{\tau}(\boldsymbol{x}) = \boldsymbol{\beta}(\boldsymbol{x})\Delta\theta_0$,

[L3] three stationary thermal conduction problem with prescribed temperature gradient $\bar{\boldsymbol{g}}^{(\alpha)}$ $(\alpha = 1, 2, 3)$.

When the finite element method is used to determine the solution of these problems a linear system of the form

$$\boldsymbol{K}\boldsymbol{U} = \boldsymbol{R} \tag{4.55}$$

has to be solved. Here $\boldsymbol{K}$ represents the modified stiffness matrix incorporating the boundary conditions, the matrix $\boldsymbol{R}$ containing the right hand side vectors and the matrix of unknowns $\boldsymbol{U}$. In order to provide an efficient solution to this system of linear equations, two different solution strategies can be applied:

[S1] Direct solution techniques.
The stiffness matrix $\boldsymbol{K}$ is factorized $\boldsymbol{K} = \boldsymbol{L}\boldsymbol{U}$ into a lower triangular matrix $\boldsymbol{L}$ and an upper triangular matrix $\boldsymbol{U}$. The solution of a linear system $\boldsymbol{K}\boldsymbol{x} = \boldsymbol{y}$ is then performed by backward-forward substitution at very low cost. However, the computation of the factorization of the matrix $\boldsymbol{K}$ is a challenging procedure and the cost in CPU time and memory is substantial. Moreover, periodic homogenization leads to a tremendous increase in the memory requirements (Fritzen and Böhlke, 2010a).

[S2] Preconditioned iterative solver.
The seven (or three) right-hand side vectors of the linear system are first evaluated. Then the stiffness matrix $\boldsymbol{K}$ of the finite element system is computed and an approximate factorization $\tilde{\boldsymbol{K}} = \tilde{\boldsymbol{L}}\tilde{\boldsymbol{U}} \approx \boldsymbol{K}$ is constructed. The approximate factorization is then used in an iterative solution technique such as the conjugate gradient method (CG; only for symmetric positive systems) or the global minimum residual method (GMRES; general non-symmetric system). Advantages of iterative solvers in the context of periodic homogenization problems were discussed by Fritzen and Böhlke (2010a). It was found that the iterative solvers - when using a sufficiently good preconditioner - can usually outperform direct solvers for a single right hand side vector. Further investigations have shown, that this does still hold true in the case of a hand-full (here: seven or three) of different right-hand side vectors. The memory requirements for these computations are significantly smaller than when using direct solution techniques. In the case of large scale computations with limited main memory it may, hence, be necessary to fall back on iterative techniques. High performance preconditioned iterative solvers are ILUPACK (`http://ilupack.tu-bs.de/`; sequential version only) or the parallel iterative package HIPS (`http://hips.gforge.inria.fr/`).

4.4.2 Approaches to computational nonlinear homogenization

The linear thermo-mechanical homogenization problems addressed in 4.2 are simplified by the fact that a limited number of reference solutions suffices to provide the exact homogenized material response. In the following the physically nonlinear behavior in a geometrically linear setting is accounted for. The local and, hence, global constitutive response are functions of the load history in this context. In particular, a path dependent behavior is observed, i.e., the final solution depends not only on the current load state given in terms of, e.g., $\bar{\varepsilon}$ (or $\bar{\sigma}$) for mechanical problems, but on the entire loading history. The generally nonlinear homogenization problem can, thus, not be solved uniquely once for all without strong additional assumptions on the geometry of the microstructure, the prescribed loading and the constitutive behavior on the microscale.

This has given rise to countless approaches in the literature (see, e.g., Kamiński, 2005; Mishnaevsky, 2007) and a universal solution is far from being established even for isothermal problems. Accounting for the classification of homogenization schemes presented in Fig. 4.5, hybrid and computational methods have major advantages over the semi-analytical approaches when it comes to restrictions due to underlying assumptions. In particular, these methods refer to exact (in the numerical sense) solutions of a set of reference problems with precise geometrical representation and in the presence of the possibly anisotropic local material response. Due to the addressed path dependency it is, however, impossible to extrapolate from a few observed load histories to different ones. Moreover, not only the final applied load determines the resulting material behavior, but also the path at which is applied. The trivial *brute force approach* in which the microstructure is fully resolved in the macroscopic problem is unacceptable with respect to computational time and efficiency.

A computational approach seeing much attention in academic applications of two-dimensional homogenization is the nested finite element method often referred to as *FE*^{2} method. It was first introduced by Renard and Marmonier (1987) and has seen a rapid development (see, for instance, Smit et al., 1998; Miehe et al., 1999; Feyel, 2003). The key idea of nested finite element approaches is to replace each integration point of the macroscopic finite element discretization with a discretization of the underlying microstructure. The strain at the integration point level then provides the boundary data for the microscopic problem. The set of all microscopic problems is then solved based on an FEM approach and the results are passed back to the structural problem in terms of the effective stress response and the tangent operators (e.g., Miehe et al., 1999). A schematic representation of the procedure is provided in Fig. 4.8. A review on FE^2 like computational homogenization methods has recently been published by Geers et al. (2010), where some of the many remaining challenges in the field have been formulated.

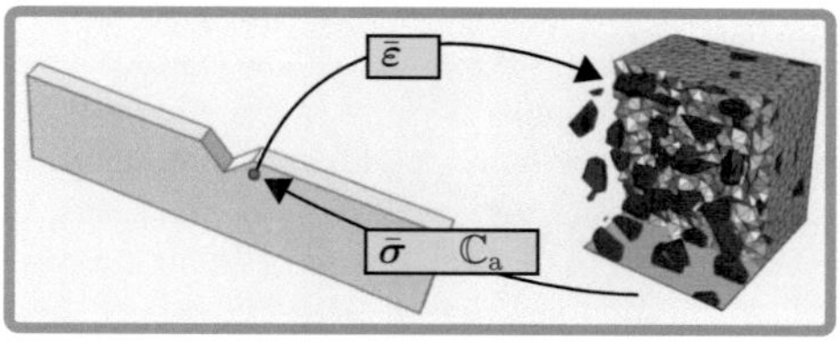

Figure 4.8: Nested finite element approach (here: two discretization levels)

The nested finite element method allows to equip the microstructured material with almost arbitrary local material behavior, i.e. no restrictions on the constitutive modeling of the microscopic constituents are induced by the method. Moreover, the geometry can exactly be captured and the results are - given a sufficient resolution of the spatial discretization on both scales - close to the real solution, if the observed micro- and macro-scale are clearly separated. These advantageous properties do not come for free. To be precise the memory and disk storage requirements are excessive and the solution of the micro problems produces tremendous computational cost. Although this cost can easily be spread over a large number of computers using parallelization techniques, it is questionable if the approach is practicable for most real world applications.

Other less sophisticated approaches start from a phenomenological model on the macroscopic scale. For example, it can be assumed that a composite consisting of an isotropic elasto-plastic matrix material and with physically and geometrically isotropic inclusions results into an effective isotropic material response of von Mises type. Using a discretization of the microstructure, a phenomenological hardening law can be identified from a hand full of reference computations on the unit cell level. Such approaches are only possible for rather specific microstructures, i.e., for simple material laws and reasonably simple geometries, e.g., isotropic arrangements of spheres. No real scale transition is performed for these methods. An example of this homogenization technique is given in section 6.3 for porous metal plasticity.

Most of the approaches based on computational methods incorporate a significant theoretical background similar to semi-analytical methods introduced in section 4.3. They build the important group of hybrid homogenization methods. The drawbacks of the semi-analytic approaches such as simplified geometry and restricted material symmetry are circumvented by hybrid techniques by replacing the respective analytic formulas by numeric computations to determine the exact (in the numerical sense) localization operators. Generally the microscopic solutions are computed only once and the solutions are stored as *eigensolutions*. The constitutive equations are then formulated in terms of a set of reduced variables such that many of the hybrid methods belong to the more general class of order-reduction techniques.

An early approach in this direction is the transformation field analysis (TFA) developed by Dvorak and Benveniste (1992); Dvorak et al. (1994a,b). In the TFA, each phase of the microstructure is divided into subdomains. Within each of these domains the inelastic strain is assumed as a piecewise constant field. The method exploits the linearity of elastic eigenstress problems to compute the local stress and strain fields by virtue to the superposition principle. Notably, the latter can for certain geometries be approximated by semi-analytical approaches, but is generally performed using the finite element method. Based on the assumed uniformity of the inelastic variables in the subdomains it is possible to determine the effective constitutive response as a function of the phase average of the stress tensor. Importantly, interactions between individual subdomains exist in terms of the induced stress and strain fields. This leads to a coupling of the evolution of the inelastic variables in different subdomains.

The TFA has seen substantial attention in the literature (see, e.g., Fish et al., 1997; Dvorak and Zhang, 2001; Carrere et al., 2004; Kruch et al., 2006; Franciosi and Berbenni,

2007) for many different material laws including elasto-visco-plasticity, crystal plasticity and damage. However, it has been recognized that the TFA may require a partitioning into sometimes numerous subdomains to not overestimate the effective nonlinear stress strain relation. This is complicated by the fact that the partitioning of the microstructure is a non-trivial task and the quality of the subdivision is hard to quantify. Having a large number of subdomains may also lead to prohibitively large nonlinear systems of equations to be solved in the time integration of the constitutive response and a large number of history variables has to be stored in the structural computations, which leads to increased memory requirements.

Based on the concept of the transformation field analysis a modification was first introduced by Michel et al. (2002) and described in detail by Michel and Suquet (2003, 2004) in terms of spatially heterogeneous inelastic strain fields. Since it it based on the TFA with additional account for the spatial non-uniformity of the inelastic strain fields, the method is referred to as *nonuniform transformation field analysis (NTFA)*. The NTFA is described in detail in chapter 5. Extensions for thermo-mechanical problems, proposals for different mode identification strategies and the implicit integration of the homogenized constitutive equations are presented. A variety of numerical examples involving the NTFA are shown in section 6.5.

Chapter 5

Non-uniform transformation field analysis

5.1 Introduction

The nonuniform transformation field analysis (NTFA) is a method for the prediction of the behavior of heterogeneous materials in the presence of physical nonlinearity but in a geometrically linear context. It extends the transformation field analysis (TFA) of Dvorak and Benveniste (1992); Dvorak et al. (1994a,b) in which phase-wise constant plastic strain fields were used in order to approximate the effective behavior of microheterogeneous materials. The NTFA was first mentioned by Michel et al. (2002) and described in detail by Michel and Suquet (2003, 2004) for materials with isotropic constitutive properties on the microscale. The key idea of the NTFA is to approximate the space-time dependency of the inelastic strain $\boldsymbol{\varepsilon}^{\mathrm{p}}(t,\boldsymbol{x})$ using a N-dimensional basis of spatially heterogeneous fields $\boldsymbol{\mu}^{(\alpha)}(\boldsymbol{x})$ and time-dependent coefficients $\xi_\alpha(t)$ according to

$$\boldsymbol{\varepsilon}^{\mathrm{p}}(t,\boldsymbol{x}) \approx \sum_{\alpha=1}^{N} \xi_\alpha(t)\boldsymbol{\mu}^{(\alpha)}(\boldsymbol{x}). \tag{5.1}$$

Thereby a sort of order reduction method is obtained and the number of unknowns is reduced to N, where N is in the order of 10. Hence, a significant reduction of the computational cost with respect to full-field simulations containing hundreds of thousands of degrees of freedom can be achieved while preserving good accuracy with respect to the macroscopic stress field. More precisely, a homogenized material law referring only to the small number of internal variables is constructed based on a micro-mechanical approach assisted by numerical computations creating the basis $\boldsymbol{\mu}^{(\alpha)}(\boldsymbol{x})$ in a machine learning algorithm.

Since its initial publication the method has undergone further developments. Fritzen and Böhlke (2010b) have reformulated some of the underlying equations and have applied the NTFA to three-dimensional problems using the finite element method. Later, Fritzen and Böhlke (2011b) showed that the method is also suitable to reproduce the phase averages of the stresses. In the latter work it was also found that the effective anisotropic linear and nonlinear material behavior of composites containing elastic particles with anisotropic morphology embedded into an elasto-plastic matrix material can be captured.

An extension of the model with respect to the underlying constitutive equations was investigated by Roussette et al. (2009), where isotropic compressible visco-plastic materials are considered. More recently, Michel and Suquet (2009) applied the method to nonlinear kinematic hardening following an extension of the Armstrong and Frederick (1966) model developed by Chaboche (1991). It should be noted that the class of generalized standard materials originally used for the NTFA does not contain the latter material. Still the NTFA was found to yield good agreement to full-field simulations.

In the following the NTFA is first related to the important field of order reduction methods. Then a generalized thermo-mechanical framework is derived and the properties of the method are examined in detail. The chosen representation imposes only rather weak constraints on the admissible constitutive behavior of the microscopic materials, i.e. it is mostly independent of the inelastic mechanisms and relies entirely on a GSM formulation of thermo-elasto-(visco-)plasticity. The weak point of the method is the recourse to postulated effective dissipation potentials found by Michel and Suquet (2003, 2004). In order to find rules for the derivation of new material models, the proposed approximations are examined theoretically. The important mode identification procedure is covered as well as the numerical implementation at different stages of the method including the mode identification process, the computation of the various coefficients and an implicit time integration procedure for the homogenized material law.

5.2 Relation to order reduction methods

In general the NTFA approach can be considered to belong to the group of order reduction algorithms usually based on the Proper Orthogonal Decomposition (POD). The POD (Karhunen, 1946; Loève, 1963) is a basic statistical tool that can extract a low dimensional basis from a set of data. It was first introduced in the field of nonlinear mechanics by Lumley (1967). The main objective of order reduction algorithms is the parametrization of the solution vector $\boldsymbol{u}^{\mathrm{h}}$ of a discrete representation of a continuum problem. Suppose $\boldsymbol{u}^{\mathrm{h}}$ is an n-dimensional real valued vector. Then an order reduction to a m-dimensional reduced vector $\boldsymbol{u}^{\mathrm{r}}$ (for $m \leq n$) can mathematically be expressed by the finite dimensional approximation

$$\boldsymbol{u}^{\mathrm{h}} = \sum_{i=1}^{m} u_i^{\mathrm{r}} \boldsymbol{v}^{(i)} + \boldsymbol{u}^{\mathrm{h}}_{\mathrm{err}}. \tag{5.2}$$

Here the vectors $\{\boldsymbol{v}^{(i)}\}$ $(i = 1, \ldots, m)$ denote a reduced basis and $\boldsymbol{u}^{\mathrm{h}}_{\mathrm{err}}$ is an error term. Often $\boldsymbol{u}^{\mathrm{h}}$ represents the nodal displacement vector of a finite element system consisting of a large number of rather simple ansatz functions that are defined piecewise on the individual elements. Neglecting the error term $\boldsymbol{u}^{\mathrm{h}}_{\mathrm{err}}$ a new low-dimensional basis of spatially rather complicated *global* shape functions can be defined. Therefore the finite element ansatz

$$u(\boldsymbol{x}, \boldsymbol{u}^{\mathrm{h}}) = \sum_{i=1}^{n} \phi_i(\boldsymbol{x}) u_i^{\mathrm{h}} \tag{5.3}$$

is introduced. For a given basis $\{\boldsymbol{v}^{(i)}\}$ one obtains the approximation

$$u(\boldsymbol{x}, \boldsymbol{u}^{\mathrm{r}}) \approx \sum_{j=1}^{m} \left(\sum_{i=1}^{n} \phi_i(\boldsymbol{x}) v_i^{(j)} \right) u_j^{\mathrm{r}} = \sum_{j=1}^{m} \tilde{\phi}_j(\boldsymbol{x}) u_j^{\mathrm{r}}. \tag{5.4}$$

As mentioned before, the new ansatz functions $\tilde{\phi}_j$ are generally defined globally instead of the local ansatz functions ϕ_i whose support is limited to the neighboring elements. The basis vectors $\boldsymbol{v}^{(i)}$ are obtained using the POD applied to a finite number of solution vectors $\boldsymbol{u}^{\mathrm{h},(i)}$ forming a sort of database for the machine learning process. The vectors $\boldsymbol{u}^{\mathrm{h},(i)}$ are usually the results of fully resolved calculations. While these approaches are usually able to improve the computational efficiency for large scale simulations, their use in the context of homogenization problems has seen few attention in the literature (Ganapathysubramanian and Zabaras, 2004; Ryckelynck and Benziane, 2010).

The parametrization of additional fields, such as the plastic strain or the vector of hardening variables, has only very recently been addressed in the hyper-reduction method (see, e.g., Ryckelynck, 2009; Ryckelynck et al., 2010). While the hyper-reduction method reduces the number of required integration points significantly over other order reduction methods, it is still required to store internal variable data at several hundred or thousand integration points. Regarding the final aim of many numerical homogenization methods, i.e. the application of the homogenized material law at each point of a macroscopic structure, the number of variables that have to be stored is, hence, exceedingly large. In fact only a few dozen internal variables can be considered to be admissible at best for three-dimensional real-world structural problems.

The advantage of the aforementioned methods is their generality with respect to the underlying physics, but this generality does not come for free. While general order reduction algorithms refer to the proper orthogonal decomposition (POD) of discrete solution vectors, the NTFA refers to actual micro-mechanical fields and its derivation incorporates the physics of the problem from the start. More specifically the formulation of the NTFA relies completely on continuum mechanical field descriptions. The underlying fields are only discretized in the actual implementation of the NTFA, whereas POD based order reduction algorithms always start on a finite dimensional real-valued vector space. Thereby the NTFA attains an improved efficiency, i.e. the number of modes N is small and the resulting homogenized material law has a sleek structure requiring only a hand full of internal variables. This is only possible by accounting for the interactions of the different fields.

In the following the elements $\boldsymbol{\mu}^{(\alpha)}$ are referred to as *inelastic modes* and the scalars ξ_α are named *mode activity coefficients* or *mode stimulation coefficients*. Due to different physical mechanisms causing plasticity in the individual constituents, a natural hypothesis imposed on the inelastic modes is (see also Michel and Suquet (2003))

[H1] The support of mode α is restricted to one constituent m_α. The domain occupied by the material labeled m is denoted Ω_m and all modes in the material are found in the index set $\mathcal{I}_m$.

The inelastic modes are identified based on numerical experiments performed on the actual microstructure of the material. Thereby, characteristic features of the heterogeneous material and the interactions due to the different local constitutive material response enter into the modes. The identification can be considered as a sort of machine learning algorithm. A detailed discussion of the mode identification procedure is presented in section 5.5. For now, it is assumed that the inelastic modes are known fields.

5.2.1 Solution to the microscopic problems

The presented investigations are based on the assumption of linearity of the stress tensor $\boldsymbol{\sigma}$ with respect to $\boldsymbol{\varepsilon} - \boldsymbol{\varepsilon}^{\mathrm{p}}$ and the temperature θ. Quasi-static conditions are assumed on the microscale. As stated in section 4.1, the balance of linear momentum on the microscopic scale (4.13) has to be satisfied. More precisely, the partial differential equation

$$\operatorname{div}\left(\mathbb{C}[\boldsymbol{\varepsilon}(t,\boldsymbol{x}) + (\theta-\theta_0)\boldsymbol{\beta}(t,\boldsymbol{x}) - \sum_{\alpha=1}^{N} \xi_\alpha(t)\boldsymbol{\mu}^{(\alpha)}(\boldsymbol{x})]\right) = \mathbf{0}, \quad \langle \boldsymbol{\varepsilon}(t,\boldsymbol{x})\rangle = \bar{\boldsymbol{\varepsilon}}(t) \quad (5.5)$$

has to be satisfied for arbitrary mode activity coefficients $\hat{\xi}(t)$, arbitrary macroscopic strains $\bar{\boldsymbol{\varepsilon}}(t)$ and for any temperature $\theta = \bar{\theta}$ in the considered setting. The solution of (5.5) is pursued in two steps: First, zero mode activity is assumed, i.e. $\xi_\alpha = 0$ $(\alpha = 1,\ldots,N)$. Thereby the seven linear thermo-elastic load cases presented in section 4.2 are recovered. The therein developed procedure can be used to compute the solution in terms of the elastic and thermal stresses $\boldsymbol{\sigma}_{\mathrm{e}}^{(\alpha)}(\boldsymbol{x}), \boldsymbol{\sigma}_\theta(\boldsymbol{x})$, the strains $\boldsymbol{\varepsilon}_{\mathrm{e}}^{(\alpha)}(\boldsymbol{x}), \boldsymbol{\varepsilon}_\theta(\boldsymbol{x})$ and the displacement fields $\boldsymbol{u}_{\mathrm{e}}^{(\alpha)}(\boldsymbol{x}), \boldsymbol{u}_\theta(\boldsymbol{x})$ for $\alpha = 1,\ldots,6$. For convenience $\langle \boldsymbol{\varepsilon}_{\mathrm{e}}^{(\gamma)}\rangle = \boldsymbol{B}^{(\gamma)}$ is chosen, with the orthonormal basis $\boldsymbol{B}^{(\gamma)}$ defined in appendix A. These fields define the localization operators under thermo-elastic loading.

In a second step, a solution of (5.5) for non-zero mode activity coefficients $\hat{\xi} \neq 0$ is constructed based on the solution of N auxiliary elastic eigenstress problems of the form

$$\operatorname{div}(\mathbb{C}[\boldsymbol{\varepsilon}_*^{(\alpha)}(\boldsymbol{x}) - \boldsymbol{\mu}^{(\alpha)}(\boldsymbol{x})]) = \mathbf{0}, \qquad \langle \boldsymbol{\varepsilon}_*^{(\alpha)} \rangle = \mathbf{0} \quad (\alpha = 1, \ldots, N). \tag{5.6}$$

The resulting stress fields are denoted $\boldsymbol{\sigma}_*^{(\alpha)}(\boldsymbol{x}) = \mathbb{C}[\boldsymbol{\varepsilon}_*^{(\alpha)}(\boldsymbol{x}) - \boldsymbol{\mu}^{(\alpha)}(\boldsymbol{x})]$ and the displacements are referred to as $\boldsymbol{u}_*^{(\alpha)}(\boldsymbol{x})$. Accounting for the linearity of the problem, a solution to (5.5) is given by

$$\boldsymbol{u}(\boldsymbol{x}, \bar{\theta}(t), \bar{\boldsymbol{\varepsilon}}(t), \hat{\xi}(t)) = \mathbb{Y}(\boldsymbol{x})[\bar{\boldsymbol{\varepsilon}}(t)] + \frac{(\bar{\theta} - \bar{\theta}_0)}{\Delta\theta_0} \boldsymbol{u}_\theta(\boldsymbol{x}) + \sum_{\alpha=1} \xi_\alpha(t) \boldsymbol{u}_*^{(\alpha)}(\boldsymbol{x}), \tag{5.7}$$

$$\boldsymbol{\varepsilon}(\boldsymbol{x}, \bar{\theta}(t), \bar{\boldsymbol{\varepsilon}}(t), \hat{\xi}(t)) = \mathbb{A}(\boldsymbol{x})[\bar{\boldsymbol{\varepsilon}}(t)] + \frac{(\bar{\theta} - \bar{\theta}_0)}{\Delta\theta_0} \boldsymbol{\varepsilon}_\theta(\boldsymbol{x}) + \sum_{\alpha=1} \xi_\alpha(t) \boldsymbol{\varepsilon}_*^{(\alpha)}(\boldsymbol{x}), \tag{5.8}$$

$$\boldsymbol{\sigma}(\boldsymbol{x}, \bar{\theta}(t), \bar{\boldsymbol{\varepsilon}}(t), \hat{\xi}(t)) = \mathbb{C}(\boldsymbol{x})\mathbb{A}(\boldsymbol{x})[\bar{\boldsymbol{\varepsilon}}(t)] + \frac{(\bar{\theta} - \bar{\theta}_0)}{\Delta\theta_0} \boldsymbol{\sigma}_\theta(\boldsymbol{x}) + \sum_{\alpha=1} \xi_\alpha(t) \boldsymbol{\sigma}_*^{(\alpha)}(\boldsymbol{x}). \tag{5.9}$$

5.2.2 Thermodynamic driving forces

To this point no constitutive assumptions except the existence of a linear thermo-elastic regime with stiffness tensor $\mathbb{C}$ and the linear geometric description of the material have been used. Taking the volume average of the stress, the effective stress tensor

$$\bar{\boldsymbol{\sigma}}(\bar{\boldsymbol{\varepsilon}}(t), \bar{\theta}(t), \hat{\xi}(t)) = \bar{\mathbb{C}}[\bar{\boldsymbol{\varepsilon}}(t)] + (\bar{\theta} - \bar{\theta}_0)\bar{\boldsymbol{\beta}} + \sum_{\alpha=1}^{N} \xi_\alpha(t) \langle \boldsymbol{\sigma}_*^{(\alpha)} \rangle \tag{5.10}$$

is a linear function of the imposed macroscopic kinematic loading $\bar{\boldsymbol{\varepsilon}}(t)$, of the macroscopic temperature $\bar{\theta}(t)$ and of the mode activity $\hat{\xi}(t)$. In order to prescribe the evolution of the latter, further assumptions are required with respect to the inelastic modes and the constitutive equations. All following observations are restricted to the class of generalized standard materials (see section 2.2.3). It is asserted that the inner variables for each of the materials are the plastic strain $\boldsymbol{\varepsilon}^{\mathrm{p}}$ and a vector of hardening variables $\hat{q}$. Additionally, a sub-class of GSM allowing for an additive decomposition of the free energy of the type

$$\psi(\theta, \boldsymbol{\varepsilon}, \boldsymbol{\varepsilon}^{\mathrm{p}}, \hat{q}) = \psi_{\mathrm{e}}(\theta, \boldsymbol{\varepsilon}, \boldsymbol{\varepsilon}^{\mathrm{p}}) + \psi_{\mathrm{k}}(\theta, \boldsymbol{\varepsilon}^{\mathrm{p}}) + \psi_{\mathrm{h}}(\theta, \hat{q}) + \psi_\theta(\theta) \tag{5.11}$$

is considered. The individual potentials are the specific strain energy ψ_{e}, the kinematic hardening potential ψ_{k}, the isotropic hardening potential ψ_{h} and a thermal energy ψ_θ. Accounting for the previously stated assumptions on the elastic behavior, the specific strain energy ψ_{e} is

$$\psi_{\mathrm{e}}(\theta, \boldsymbol{\varepsilon}, \boldsymbol{\varepsilon}^{\mathrm{p}}) = \frac{1}{2\varrho}(\boldsymbol{\varepsilon} - \boldsymbol{\varepsilon}^{\mathrm{p}}) \cdot \mathbb{C}[\boldsymbol{\varepsilon} - \boldsymbol{\varepsilon}^{\mathrm{p}}] + \frac{1}{\varrho}(\boldsymbol{\varepsilon} - \boldsymbol{\varepsilon}^{\mathrm{p}}) \cdot \boldsymbol{\beta}(\theta - \theta_0). \tag{5.12}$$

Then the macroscopic free energy density $\bar{w}$ per unit volume of the heterogeneous material is defined by the volume average

$$\bar{w} = \langle \varrho\psi \rangle = \langle \varrho\psi_{\mathrm{e}}(\theta, \boldsymbol{\varepsilon}, \boldsymbol{\varepsilon}^{\mathrm{p}}) \rangle + \langle \varrho\psi_{\mathrm{k}}(\theta, \boldsymbol{\varepsilon}^{\mathrm{p}}) \rangle + \langle \varrho\psi_{\mathrm{h}}(\theta, \hat{q}) \rangle + \langle \varrho\psi_{\theta}(\theta) \rangle. \tag{5.13}$$

Due to the additivity of the energy this volume average is uniquely defined. New variables at the macroscopic (or structural) level are the macroscopic strain $\bar{\boldsymbol{\varepsilon}}$, the macroscopic temperature $\bar{\theta}$ and the mode activity vector $\hat{\xi}$. Additionally, the hardening variables $\hat{q}$ need to be specified. In their approach to the topic, Michel and Suquet assumed the latter to be constituent-wise constant quantities.

[H2] The vector of hardening variables $\hat{q}$ is constant within each constituent.

Hence, the hardening variables can be identified by their effective value $\hat{\bar{q}}^{(m)}$ $(1 \leq m \leq M)$. As stated earlier, the macroscopic and the microscopic temperature need not be distinguished due to the assumed separation of scales, i.e., $\theta = \bar{\theta}$ holds. For a set of M different constituents found in the microstructure the formula

$$\bar{w} = \langle \varrho\psi_{\mathrm{e}}(\bar{\theta}, \boldsymbol{\varepsilon}, \boldsymbol{\varepsilon}^{\mathrm{p}}) \rangle + \langle \varrho\psi_{\mathrm{k}}(\bar{\theta}, \boldsymbol{\varepsilon}^{\mathrm{p}}) \rangle + \left(\sum_{m=1}^{M} c^{(m)} \varrho^{(m)} \psi_{\mathrm{h}}^{(m)}(\bar{\theta}, \hat{\bar{q}}^{(m)}) \right) + \sum_{m=1}^{M} c^{(m)} \varrho^{(m)} \psi_{\theta}^{(m)}(\bar{\theta}) \tag{5.14}$$

is attained. The superscript index $^{(m)}$ at ψ_{h} and ψ_{θ} denotes the phase dependency of these functions and $c^{(m)}$, $\varrho^{(m)}$ are the volume fraction and the mass density of material m, respectively. Then the thermodynamic driving forces with respect to the mode activity coefficients are $(\alpha = 1, \ldots, N)$

$$\tau_{\alpha} = -\frac{\partial \bar{w}}{\partial \xi_{\alpha}} = -\left\langle \varrho \frac{\partial \psi}{\partial \xi_{\alpha}} \right\rangle = -\left\langle \frac{\partial \varrho\psi_{\mathrm{e}}(\bar{\theta}, \boldsymbol{\varepsilon}, \boldsymbol{\varepsilon}^{\mathrm{p}})}{\partial \xi_{\alpha}} + \frac{\partial \varrho\psi_{\mathrm{k}}(\bar{\theta}, \boldsymbol{\varepsilon}^{\mathrm{p}})}{\partial \xi_{\alpha}} \right\rangle. \tag{5.15}$$

Making use of of the identity

$$\frac{\partial \bullet (\boldsymbol{\varepsilon}, \boldsymbol{\varepsilon}^{\mathrm{p}}, \ldots)}{\partial \xi_{\alpha}} = \frac{\partial \bullet (\boldsymbol{\varepsilon}, \boldsymbol{\varepsilon}^{\mathrm{p}}, \ldots)}{\partial \boldsymbol{\varepsilon}} \cdot \boldsymbol{\varepsilon}_{*}^{(\alpha)} + \frac{\partial \bullet (\boldsymbol{\varepsilon}, \boldsymbol{\varepsilon}^{\mathrm{p}}, \ldots)}{\partial \boldsymbol{\varepsilon}^{\mathrm{p}}} \cdot \boldsymbol{\mu}^{(\alpha)}, \tag{5.16}$$

the individual terms in (5.15) can be replaced and one gets

$$\tau_{\alpha} = \langle \boldsymbol{\sigma} \cdot (\boldsymbol{\mu}^{(\alpha)} - \boldsymbol{\varepsilon}_{*}^{(\alpha)}) \rangle + \left\langle \frac{\partial \varrho\psi_{\mathrm{k}}}{\partial \boldsymbol{\varepsilon}^{\mathrm{p}}} \cdot \boldsymbol{\mu}^{(\alpha)} \right\rangle \quad (\alpha = 1, \ldots, N). \tag{5.17}$$

Limiting attention to linear kinematic hardening the potential ψ_{k} can be written as

$$\psi_{\mathrm{k}} = \frac{1}{2\varrho} k(\bar{\theta}) \boldsymbol{\varepsilon}^{\mathrm{p}}(\hat{\xi}) \cdot \mathbb{K}[\boldsymbol{\varepsilon}^{\mathrm{p}}(\hat{\xi})], \tag{5.18}$$

where $k(\bar{\theta})$ is a non-dimensional function expressing the thermal sensitivity of the kinematic hardening and $\mathbb{K}$ is a semi-positive definite, major and minor symmetric fourth-order tensor denoting the hardening metric. Both are asserted constant within each constituent, i.e.,

$\mathbb{K} = \mathbb{K}^{(m)}$ and $k(\bar{\theta}) = k^{(m)}(\bar{\theta})$ in Ω_m $(m = 1, \ldots, M)$. With the index set $\mathcal{I}_{m_\alpha}$ one is able to rewrite (5.17) as $\quad (\alpha = 1, \ldots, N)$

$$\tau_\alpha = \langle \mathbb{C}\mathbb{A}[\bar{\varepsilon}] \cdot (\boldsymbol{\mu}^{(\alpha)} - \boldsymbol{\varepsilon}_*^{(\alpha)}) \rangle + \frac{1}{\Delta\theta_0}(\bar{\theta} - \bar{\theta}_0)\langle \boldsymbol{\sigma}_\theta \cdot (\boldsymbol{\mu}^{(\alpha)} - \boldsymbol{\varepsilon}_*^{(\alpha)}) \rangle$$
$$+ \sum_{\beta=1}^{N} \xi_\beta \langle \boldsymbol{\sigma}_*^{(\beta)} \cdot (\boldsymbol{\mu}^{(\alpha)} - \boldsymbol{\varepsilon}_*^{(\alpha)}) \rangle - k^{(m_\alpha)}(\bar{\theta}) \sum_{\gamma \in \mathcal{I}_{m_\alpha}} \xi_\gamma \langle \boldsymbol{\mu}^{(\gamma)} \cdot \mathbb{K}^{(m_\alpha)}[\boldsymbol{\mu}^{(\alpha)}] \rangle. \quad (5.19)$$

The previous expression can be reorganized into a more convenient matrix vector notation. Therefore the matrices $(\alpha, \beta = 1, \ldots, N; \gamma = 1, \ldots, 6)$

$$A_{\alpha\gamma} = \langle \mathbb{C}\mathbb{A}[\boldsymbol{B}^{(\gamma)}] \cdot (\boldsymbol{\mu}^{(\alpha)} - \boldsymbol{\varepsilon}_*^{(\alpha)}) \rangle = \langle \mathbb{A}^\mathsf{T}\mathbb{C}[\boldsymbol{\mu}^{(\alpha)} - \boldsymbol{\varepsilon}_*^{(\alpha)}] \rangle \cdot \boldsymbol{B}^{(\gamma)}, \quad (5.20)$$
$$D^0_{\alpha\beta} = \langle (\boldsymbol{\mu}^{(\alpha)} - \boldsymbol{\varepsilon}_*^{(\alpha)}) \cdot \boldsymbol{\sigma}_*^{(\beta)} \rangle, \quad (5.21)$$
$$K^0_{\alpha\beta} = -\langle \boldsymbol{\mu}^{(\alpha)} \cdot \mathbb{K}^{(m_\alpha)}[\boldsymbol{\mu}^{(\beta)}] \rangle, \quad K_{\alpha\beta}(\bar{\theta}) = k^{(m_\alpha)}(\bar{\theta}) K^0_{\alpha\beta} \quad (5.22)$$
$$\hat{D} = \hat{D}^0 + \hat{K}(\bar{\theta}), \quad (5.23)$$
$$R_\alpha = \frac{1}{\Delta\theta_0}\langle \boldsymbol{\sigma}_\theta \cdot (\boldsymbol{\mu}^{(\alpha)} - \boldsymbol{\varepsilon}_*^{(\alpha)}) \rangle, \quad (5.24)$$

are defined. Then the vector of thermodynamic driving forces conjugate to $\hat{\xi}$ is

$$\hat{\tau} = \hat{A}\hat{\bar{\varepsilon}} + \hat{D}(\bar{\theta})\hat{\xi} + (\bar{\theta} - \bar{\theta}_0)\hat{R} = \hat{A}\hat{\bar{\varepsilon}} + (\hat{D}^0 + \hat{K}(\bar{\theta}))\hat{\xi} + (\bar{\theta} - \bar{\theta}_0)\hat{R}. \quad (5.25)$$

Additionally, for each inelastic constituent the generalized vector of forces conjugated to the hardening variables $\hat{\bar{q}}^{(m)}$ $(m = 1, \ldots, M)$ computes to

$$\hat{\bar{r}}^{(m)} = -c^{(m)}\varrho^{(m)}\frac{\partial\psi_\mathrm{h}^{(m)}(\bar{\theta}, \hat{\bar{q}}^{(m)})}{\partial\hat{\bar{q}}^{(m)}}. \quad (5.26)$$

In previous works dealing with the NTFA (see, e.g., Michel and Suquet, 2003, 2004; Fritzen and Böhlke, 2010b) the derivation of the vector $\hat{\tau}$ of hardening variables was not based on the partial derivative of the effective free energy, but on the homogenization of the local dissipation. For the isothermal case without kinematic hardening the procedure yields

$$\tau_\alpha = \langle \mathbb{A}^\mathsf{T}\mathbb{C}[\boldsymbol{\mu}^{(\alpha)}] \rangle \cdot \bar{\varepsilon} + \sum_{\beta=1}^{N} \xi_\beta \langle \boldsymbol{\mu}^{(\alpha)} \cdot \boldsymbol{\sigma}_*^{(\beta)} \rangle. \quad (5.27)$$

The differences in terms of the missing fluctuations $\boldsymbol{\varepsilon}_*^{(\alpha)}$ of the strain tensor are noteworthy. In particular the homogenized free energy contains already information on the solution of the eigenstress problem induced by the plastic deformation. Homogenizing directly the local dissipation as done by Michel and Suquet (2003), these terms do not enter. Accounting for the Hill Mandel condition (Hill, 1963)

$$\langle \boldsymbol{\sigma} \cdot \boldsymbol{\varepsilon} \rangle = \langle \boldsymbol{\sigma} \rangle \cdot \langle \boldsymbol{\varepsilon} \rangle \quad (5.28)$$

for any statically admissible stress field $\boldsymbol{\sigma}$ and kinematically admissible strain field $\boldsymbol{\varepsilon}$, it can be shown that the formulations (5.25) and (5.27) are identical for isothermal problems without kinematic hardening. In order to show this, the representation $(\gamma = 1, \ldots, 6)$

$$\mathbb{A} = \sum_{\gamma=1}^{6} \boldsymbol{\varepsilon}_{\mathrm{e}}^{(\gamma)} \otimes \boldsymbol{B}^{(\gamma)}, \qquad \boldsymbol{\sigma}_{\mathrm{e}}^{(\gamma)} = \mathbb{C}[\boldsymbol{\varepsilon}_{\mathrm{e}}^{(\gamma)}], \qquad \langle \boldsymbol{\varepsilon}_{\mathrm{e}}^{(\gamma)} \rangle = \boldsymbol{B}^{(\gamma)}, \tag{5.29}$$

of the strain localization operator of the linear-elastic material is used. Then the identity

$$\langle \boldsymbol{\sigma}_*^{(\alpha)} \cdot \boldsymbol{\varepsilon}_*^{(\beta)} \rangle = 0 \qquad (\alpha, \beta = 1, \ldots, N) \tag{5.30}$$

holds due to $\langle \boldsymbol{\varepsilon}_*^{(\beta)} \rangle = \mathbf{0}$ and due to the static admissibility of $\boldsymbol{\sigma}_*^{(\alpha)}$. Further, the representation (5.29) is used to find the identity

$$\langle \mathbb{A}^{\mathsf{T}} \mathbb{C}[\boldsymbol{\varepsilon}_*^{(\alpha)}] \cdot \bar{\boldsymbol{\varepsilon}} \rangle = \sum_{\gamma=1}^{6} \bar{\varepsilon}_\gamma \langle \boldsymbol{\varepsilon}_*^{(\alpha)} \cdot \boldsymbol{\sigma}_{\mathrm{e}}^{(\gamma)} \rangle = \sum_{\gamma=1}^{6} \bar{\varepsilon}_\gamma \langle \boldsymbol{\varepsilon}_*^{(\alpha)} \rangle \cdot \langle \boldsymbol{\sigma}_{\mathrm{e}}^{(\gamma)} \rangle = 0. \tag{5.31}$$

Based on the previous results equation (5.27) can be reformulated into

$$\tau_\alpha = \langle \mathbb{A}^{\mathsf{T}} \mathbb{C}[\boldsymbol{\mu}^{(\alpha)} - \boldsymbol{\varepsilon}_*^{(\alpha)}] \rangle \cdot \bar{\boldsymbol{\varepsilon}} + \sum_{\beta=1}^{N} \xi_\beta \langle (\boldsymbol{\mu}^{(\alpha)} - \boldsymbol{\varepsilon}_*^{(\alpha)}) \cdot \boldsymbol{\sigma}_*^{(\beta)} \rangle. \tag{5.32}$$

This formulation is identical to the isothermal case without kinematic hardening in (5.25). In fact, the consideration of the linear kinematic hardening is identical in both approaches and does not alter the result. The thermodynamic driving forces computed based on the two different approaches are, hence, equivalent.

Note that to this point the structure of the potential (5.11) is the only assumption imposed on the inelastic constitutive behavior of the material. It allows for a variety of hardening laws without any limitation on the actual choice of $\boldsymbol{\mu}^{(\alpha)}$ being enforced to this point.

5.3 Analysis of the system matrices

Taking the representation (5.10) and rewriting it in terms of a vector-matrix representation with respect to the orthonormal basis $\boldsymbol{B}^{(\alpha)}$, one gets

$$\hat{\bar{\sigma}} = \hat{\bar{C}} \hat{\bar{\varepsilon}} + (\bar{\theta} - \bar{\theta}_0) \hat{\bar{\beta}} + \hat{\rho} \hat{\xi}, \tag{5.33}$$

where the Matrix $\hat{\rho}$ maps the mode activity onto the effective stress components. The individual entries of $\hat{\rho}$ are

$$(\hat{\rho})_{\gamma\alpha} = \boldsymbol{B}_\gamma \cdot \langle \boldsymbol{\sigma}_*^{(\alpha)} \rangle \qquad (\alpha = 1, \ldots, N; \gamma = 1, \ldots, 6). \tag{5.34}$$

Taking further into account the three matrices $\hat{A}$, $\hat{D}^0$ and $\hat{K}^0$ defining the thermodynamic conjugate forces, a total of four different system matrices emerge from the extension of the thermo-elastic material behavior through the NTFA. The components of $\hat{A}$ and $\hat{D}$ are recalled for completeness (with $\alpha, \beta = 1, \dots, N; \gamma = 1, \dots, 6$):

$$(\hat{A})_{\alpha\gamma} = \langle \mathbb{A}^{\mathsf{T}} \mathbb{C}[\boldsymbol{\mu}^{(\alpha)} - \boldsymbol{\varepsilon}_*^{(\alpha)}] \rangle \cdot \boldsymbol{B}^{(\gamma)}, \tag{5.35}$$

$$(\hat{D}^0)_{\alpha\beta} = \langle (\boldsymbol{\mu}^{(\alpha)} - \boldsymbol{\varepsilon}_*^{(\alpha)}) \cdot \boldsymbol{\sigma}_*^{(\beta)} \rangle, \tag{5.36}$$

$$(\hat{K}^0)_{\alpha\beta} = -\langle \boldsymbol{\mu}^{(\alpha)} \cdot \mathbb{K}^{(m_\alpha)}[\boldsymbol{\mu}^{(\beta)}] \rangle. \tag{5.37}$$

In the following the properties of the four matrices are examined.

Matrix $\hat{A}$. In order to simplify (5.35) the representation (5.29) of the localization operator $\mathbb{A}$ of the elastic microheterogeneous material is used (see also section 4.4.1). Further, the identity

$$\mathbb{C}[\boldsymbol{\mu}^{(\alpha)} - \boldsymbol{\varepsilon}_*^{(\alpha)}] = -\boldsymbol{\sigma}_*^{(\alpha)} \tag{5.38}$$

is substituted. Based on (4.47) and the statical admissibility of $\boldsymbol{\sigma}_*^{(\alpha)}$ the component $(\hat{A})_{\alpha\gamma}$ becomes

$$(\hat{A})_{\alpha\gamma} = \langle \mathbb{A}^{\mathsf{T}} \mathbb{C}[\boldsymbol{\mu}^{(\alpha)} - \boldsymbol{\varepsilon}_*^{(\alpha)}] \rangle \cdot \boldsymbol{B}^{(\gamma)} = -\langle \mathbb{A}^{\mathsf{T}}[\boldsymbol{\sigma}_*^{(\alpha)}] \rangle \cdot \boldsymbol{B}^{(\gamma)} = -\langle \boldsymbol{\sigma}_*^{(\alpha)} \rangle \cdot \boldsymbol{B}^{(\gamma)}. \tag{5.39}$$

Hence, the matrix $\hat{A}$ represents the negative value of the effective stresses induced by the inelastic modes. It has to be emphasized that the components of $\hat{A}$ do not express phase averages, but averages taken on the entire unit cell. Moreover, it is found that $\hat{A}$ is equivalent to the matrix $\hat{\rho}$ due to $(\alpha = 1, \dots, N; \gamma = 1, \dots, 6)$

$$(\hat{A})_{\alpha\gamma} = -\langle \boldsymbol{\sigma}_*^{(\alpha)} \rangle \cdot \boldsymbol{B}^{(\gamma)} = -(\hat{\rho})_{\gamma\alpha} \quad \Rightarrow \quad \hat{A} = -\hat{\rho}^{\mathsf{T}}. \tag{5.40}$$

Consequently, the number of (independent) system matrices is reduced to three: the matrix $\hat{A}$ containing the effective induced stresses and the matrices $\hat{D}^0$, $\hat{K}^0$ containing the mode interaction coefficients.

Effective kinematic hardening metric $\hat{K}^0$. The matrix $\hat{K}^0$ is symmetric due to the symmetry of $\mathbb{K}$. Further, the positive semi-definite nature of $\mathbb{K}$ implies that $\hat{K}^0$ is a negative semi-definite matrix. Due to $k(\bar{\theta}) \geq 0$ this implies the negative semi-definiteness of $\hat{K}(\bar{\theta})$.

Mode interaction matrix $\hat{D}^0$. Based on $\boldsymbol{\sigma}_*^{(\beta)} = \mathbb{C}[\boldsymbol{\varepsilon}_*^{(\beta)} - \boldsymbol{\mu}^{(\beta)}]$ and by introduction of the compliance tensor $\mathbb{S} = \mathbb{C}^{-1}$, the components of the matrix $\hat{D}^0$ are

$$(\hat{D}^0)_{\alpha\beta} = \langle (\boldsymbol{\mu}^{(\alpha)} - \boldsymbol{\varepsilon}_*^{(\alpha)}) \cdot \boldsymbol{\sigma}_*^{(\beta)} \rangle = -\langle \boldsymbol{\sigma}_*^{(\alpha)} \cdot \mathbb{S}[\boldsymbol{\sigma}_*^{(\beta)}] \rangle \quad (\alpha, \beta = 1, \dots, N). \tag{5.41}$$

Two conclusions are directly possible:

- The coefficient matrix is symmetric, i.e., $(\hat{D}^0)_{\alpha\beta} = (\hat{D}^0)_{\beta\alpha}$.

- The matrix $\hat{D}^0$ is negative definite, if the stress fields $\{\boldsymbol{\sigma}_*^{(\alpha)}\}$ are linearly independent. To proof the latter a vector $\hat{y} \in \mathbb{R}^N \setminus \{\hat{0}\}$ is considered. Then the product

$$\hat{y} \cdot \hat{D}^0 \hat{y} = -\Big\langle \underbrace{\big(\sum_{\alpha=1}^{N} y_\alpha \boldsymbol{\sigma}_*^{(\alpha)}\big)}_{=\boldsymbol{\sigma}_y} \cdot \mathbb{S}[\big(\sum_{\beta=1}^{N} y_\beta \boldsymbol{\sigma}_*^{(\beta)}\big)]\Big\rangle \tag{5.42}$$

has to be smaller than 0 for any choice $\hat{y}$. Given the linear independence of the fields $\boldsymbol{\sigma}_*^{(\alpha)}$, the stress tensor $\boldsymbol{\sigma}_y$ is non-zero in a measurable sub-domain of Ω. Together with the positive definiteness of the compliance tensor $\mathbb{S}$ the result

$$\hat{y} \cdot \hat{D}^0 \hat{y} = -\langle \boldsymbol{\sigma}_y \cdot \mathbb{S}[\boldsymbol{\sigma}_y] \rangle < 0 \tag{5.43}$$

is immediately recovered. Due to the arbitrariness of $\hat{y}$, the matrix $\hat{D}^0$ is symmetric negative-definite.

Combining the results obtained for $\hat{K}(\bar{\theta})$ and $\hat{D}^0$, the symmetry and the negative definiteness of $\hat{D}(\bar{\theta}) = \hat{K}(\bar{\theta}) + \hat{D}^0$ is proven.

5.4 Evolution of mode activity

5.4.1 Existing approaches

The previous results allow for the computation of the local fields given the vector of mode stimulation coefficients $\hat{\xi}$, the macroscopic strain $\bar{\varepsilon}$ and the temperature $\bar{\theta}$. For an undeformed stress-free material the vector $\hat{\xi}$ is assumed zero. Under loading the effective material behavior depends on the evolution of the components of $\hat{\xi}$, which remains to be specified. For a classical (single-scale) GSM the evolution of the internal variables depends on the vector of thermodynamic driving forces in terms of the dual dissipation potential ϕ^* (see section 2.2.3). Since the dissipation is an additive quantity (Suquet, 1985a), the exact homogenized potential $\bar{\phi}^*$ is required as a function of the macroscopic thermodynamic conjugates $\hat{\tau}, \hat{r}$, in order to determine the evolution of the mode activity. Unfortunately, $\bar{\phi}^*(\hat{\tau}, \hat{r})$ can usually not be determined exactly.

In previous studies of the NTFA, isotropic components with plastically incompressible behavior have been studied (Michel and Suquet, 2003, 2004; Fritzen and Böhlke, 2010b). An extension to isotropic rate dependent compressible materials was first formulated by (Roussette et al., 2009). In the following attention is limited to the case of rate dependent and rate independent incompressible plasticity of von Mises type. A single scalar hardening variable q resembling the accumulated equivalent plastic strain is assumed. Then $r = \varrho \partial_q \psi$ denotes an isotropic hardening stress and $\boldsymbol{\Xi} = K(\theta)\boldsymbol{\varepsilon}^{\mathrm{p}}$ is a back-stress due to linear kinematic hardening. For convenience the thermodynamic conjugate force $\mathcal{A}$ to the plastic strain and the derived equivalent stress $\mathcal{A}_{\text{eq}}$ are written as

$$\mathcal{A} = \boldsymbol{\sigma}' - \boldsymbol{\Xi}, \qquad \mathcal{A}_{\text{eq}} = \sqrt{\frac{3}{2}} \|\mathcal{A}\|_2. \tag{5.44}$$

Local material behavior. In the rate independent case the evolution of the local inelastic variables follows from

$$\dot{\boldsymbol{\varepsilon}}^{\text{p}} = \dot{\lambda} \frac{\partial \varphi}{\partial \mathcal{A}}, \qquad \dot{q} = \dot{\lambda} \frac{\partial \varphi}{\partial r} \qquad \dot{\lambda} \geq 0, \qquad \dot{\lambda} \varphi = 0, \tag{5.45}$$

where φ is a function describing the yield surface of the material via

$$\varphi(\mathcal{A}_{\text{eq}}, r) = \mathcal{A}_{\text{eq}} - (\sigma_{\text{F0}} + r(q)) \leq 0. \tag{5.46}$$

Consequently, the rates of the inelastic variables are

$$\dot{\boldsymbol{\varepsilon}}^{\text{p}} = \sqrt{\frac{3}{2}} \dot{\lambda} \frac{\boldsymbol{\sigma}' - \boldsymbol{\Xi}}{\|\boldsymbol{\sigma}' - \boldsymbol{\Xi}\|_2} = \frac{3}{2} \dot{\lambda} \frac{\mathcal{A}}{\mathcal{A}_{\text{eq}}}, \qquad \dot{q} = \dot{\lambda}. \tag{5.47}$$

The dual dissipation potential ϕ^* is a non-smooth and unbounded function (see section 2.2.3). Due to the latter it is impossible to determine the volume average of ϕ^* exactly. Moreover, the potential is, in its current form, not expressed in terms of the vector $\hat{\tau}$ denoting the thermodynamic conjugates to the mode activity.

For a rate dependent material the evolution equation for the inelastic variables is derived from

$$\phi^*(\mathcal{A}_{\text{eq}}, r) = \frac{\dot{\gamma}_0 \sigma_{\text{D}}}{n+1} \left(\frac{\max\{0, \mathcal{A}_{\text{eq}} - (\sigma_{\text{F0}} + r)\}}{\sigma_{\text{D}}} \right)^{n+1}, \tag{5.48}$$

with the drag stress σ_{D}, the reference rate $\dot{\gamma}_0$ and the viscosity exponent $n \geq 1$. They are

$$\dot{\boldsymbol{\varepsilon}}^{\text{p}} = \frac{\partial \phi^*}{\partial \mathcal{A}} = \sqrt{\frac{3}{2}} \dot{\lambda} \frac{\boldsymbol{\sigma}' - \boldsymbol{\Xi}}{\|\boldsymbol{\sigma}' - \boldsymbol{\Xi}\|_2} = \frac{3}{2} \dot{\lambda} \frac{\mathcal{A}}{\mathcal{A}_{\text{eq}}}, \qquad \dot{q} = -\frac{\partial \phi^*}{\partial r} = \dot{\lambda}, \tag{5.49}$$

with

$$\dot{\lambda} = \dot{\gamma}_0 \left(\frac{\max\{0, \mathcal{A}_{\text{eq}} - (\sigma_{\text{F0}} + r)\}}{\sigma_{\text{D}}} \right)^n. \tag{5.50}$$

The potential ϕ^* of visco-plastic constituents is a two-times continuously differentiable. It is, however, impossible to compute the volume average $\bar{\phi}^*$ as a function of $\hat{\tau}$ due to the maximum condition. A special case of (5.49) is a purely viscous material which is attained for $\sigma_{\text{F0}} = r = 0$.

While a general solution to the homogenization of the evolution law of the internal variables is an open question of scientific significance, approximations are possible for the described class of materials based on an approximation of $\bar{\phi}^*$. The following investigations are based on a proposal due to Michel and Suquet (2003, 2004) which has been the basis for subsequent proposals. They imposed the following requirements on the modes:

[M1] The modes are L^2 orthogonal, i.e., $\langle \boldsymbol{\mu}^{(\alpha)} \cdot \boldsymbol{\mu}^{(\beta)} \rangle = 0$ $(\alpha \neq \beta)$.

[M2] The modes are normalized $\langle \|\boldsymbol{\mu}^{(\alpha)}\|_2 \rangle = \sqrt{3/2}$.

Based on these assertions and the hypotheses H1 and H2, the uncoupled and the coupled model were proposed. In both models the comparison stress $\mathcal{A}_{\rm eq} = \sqrt{3/2}\|\boldsymbol{\sigma} - \boldsymbol{\Xi}\|_2$ is replaced by an approximation $\bar{\mathcal{A}}_{\rm eq}$. The thereby modified evolution equations are then differentiated with respect to the conjugate forces to provide the evolution for the vector $\hat{\xi}$.

In the uncoupled model, each mode is associated with an individual effective dissipation potential $\bar{\phi}^*_\alpha$. Within each of these potentials the variable $\mathcal{A}_{\rm eq}$ is approximated with the absolute value of the component τ_α. Consequently, each mode then has an individual (constituent-wise constant) hardening variable q_α and an individual potential $\bar{\phi}^*_\alpha$.

In the coupled model, the variable $\mathcal{A}_{\rm eq}$ is approximated by the euclidean norm of $\hat{\tau}$ in the yield criterion for each constituent. The corresponding rate potential is denoted $\bar{\phi}^{*,(m)}$. The two models are compared in Fig. 5.1.

uncoupled model	coupled model
N potentials $\bar{\phi}^*_\alpha \equiv \phi^*(\bar{\mathcal{A}}_{\rm eq,\alpha}, \bar{r}_\alpha)$	M potentials $\bar{\phi}^{*,(m)} \equiv \phi^*(\bar{\mathcal{A}}^{(m)}_{\rm eq}, \bar{r}^{(m)})$
$\bar{\mathcal{A}}_{\rm eq,\alpha} = \lvert\tau_\alpha\rvert$	$\bar{\mathcal{A}}^{(m)}_{\rm eq} = \|\hat{\tau}^{(m)}\|_2$
N hardening variables $\bar{q}_\alpha$	M hardening variables $\bar{q}^{(m)}$
N isotropic hardening stresses $\bar{r}_\alpha$	M isotropic hardening stresses $\bar{r}^{(m)}$

Figure 5.1: Formal comparison of the uncoupled model and the coupled model of the NTFA cf. Michel and Suquet (2003, 2004)

5.4.2 Verification and falsification

In order to judge on the quality of the previously introduced approximations, a simple reference problem is examined analytically. A minimum requirement is imposed on any approximation in the sense that the approximation $\bar{\phi}^*$ of the effective rate potential can reproduce the evolution equation of a homogeneous reference material *exactly*. Only then it is expected that a model can replicate the response of a heterogeneous material to a sufficient extent.

In order to allow for these observations, a homogeneous reference volume Ω made of a single inelastic constituent and without voids is considered. For such a homogeneous material all fields are homogeneous, even in the presence of inelasticity. To be more precise, thermodynamic material stability is assumed. Then the corresponding incremental Gibb's potential (see, e.g., Hackl, 1996; Ortiz and Repetto, 1999) is (at least) a quasi-convex function of the strain*. Additionally, it is asserted that a sufficient number of virtual experiments is conducted with the suggested homogeneous reference material and that the thereby identified modes span the space of inelastic deformation permitted by the material.

In fact, in the context of von Mises type plasticity, the homogeneity of the material and, thereby, all fields reduces the number of the required experiments to five linearly independent deviatoric loadings. Incorporating the mode restrictions M1, M2 and using the approximated evolution law for the mode coefficients the material response of the material as predicted by the NTFA can be computed. It can then be compared to the known, i.e., exact, solution of the homogeneous material, see Fig. 5.2. The imposed minimum requirement is equivalent to requiring $\bar{\phi}^* \equiv \phi^*$ for any thermodynamic state of the material. For simplicity no kinematic hardening is considered and an isothermal set-up is examined.

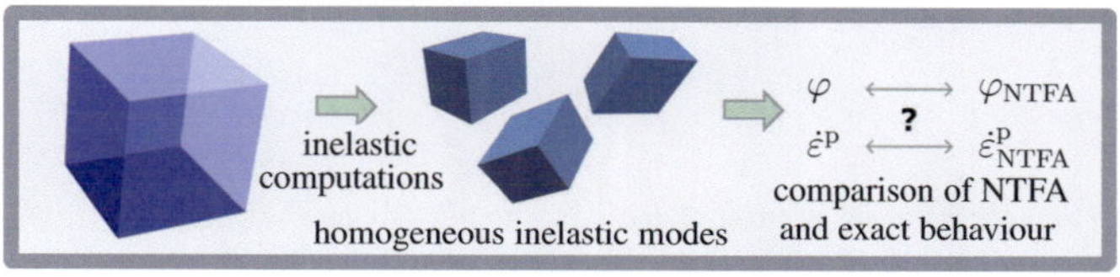

Figure 5.2: Scheme of a verification procedure for the inelastic behavior of the homogenized NTFA response of a homogeneous comparison material

Due to the homogeneity of all involved fields, the local value and the average value need not be distinguished. In particular all fluctuation fields vanish. By virtue to the required L^2_Ω-orthogonality of the modes (M1), an arbitrary orthonormal basis $\boldsymbol{B}'^{(\alpha)}$ (as e.g. in append A) of the five-dimensional space of deviatoric tensors can be chosen for the five modes. Accounting for the imposed mode normalization (M2) one obtains

$$\boldsymbol{\mu}^{(\alpha)} = \sqrt{\frac{3}{2}} \boldsymbol{B}'^{(\alpha)}. \tag{5.51}$$

Further, the tensor $\boldsymbol{B}^\circ = \frac{\sqrt{3}}{3}\boldsymbol{I}$ augments the basis $\{\boldsymbol{B}'^{(\alpha)}\}_{\alpha=1,\dots,5}$ to span the space of all symmetric tensors. Each strain tensor $\boldsymbol{\varepsilon}$ and stress tensor $\boldsymbol{\sigma}$ then has a vector representation

$$\boldsymbol{\varepsilon} = \varepsilon^\circ \boldsymbol{B}^\circ + \sum_{\alpha=1}^{5} \varepsilon'_\alpha \boldsymbol{B}'^{(\alpha)}, \qquad \varepsilon^\circ = \boldsymbol{\varepsilon} \cdot \boldsymbol{B}^\circ, \qquad \varepsilon'_\alpha = \boldsymbol{\varepsilon} \cdot \boldsymbol{B}^{(\alpha)}, \tag{5.52}$$

$$\boldsymbol{\sigma} = \sigma^\circ \boldsymbol{B}^\circ + \sum_{\alpha=1}^{5} \sigma_\alpha \boldsymbol{B}'^{(\alpha)}, \qquad \sigma^\circ = \boldsymbol{\sigma} \cdot \boldsymbol{B}^\circ, \qquad \sigma'_\alpha = \boldsymbol{\sigma} \cdot \boldsymbol{B}^{(\alpha)}. \tag{5.53}$$

*Otherwise the material response defines an ill-posed problem with non-unique (if any) solution and the effort for solution of such problems is, even without the additional burden of a scale transition, a computational adventure. The examination of the different convexity requirements is subject of an extensive literature, e.g., Ball (1977); Mielke (2004) to mention only a few original contributions.

Now the stress tensor as a function of the strain and the five mode stimulation coefficients is considered. The induced stress and strain fields are

$$\boldsymbol{\varepsilon}_*^{(\alpha)} = \mathbf{0}, \qquad \boldsymbol{\sigma}_*^{(\alpha)} = -\sqrt{\frac{3}{2}}\,\mathbb{C}[\boldsymbol{B}'^{(\alpha)}]. \tag{5.54}$$

Assembling all quantities and accounting for isotropic elasticity with the shear modulus G and the bulk modulus K gives

$$\boldsymbol{\sigma} = \sqrt{3}K\varepsilon^{\circ}\boldsymbol{I} + 2G\sum_{\alpha=1}^{5}\boldsymbol{B}'^{(\alpha)}\left(\varepsilon'_{\alpha} - \sqrt{\frac{3}{2}}\xi_{\alpha}\right). \tag{5.55}$$

Based on the latter representation, the conjugate forces compute to

$$\tau_{\alpha} = \boldsymbol{\mu}^{(\alpha)}\cdot\boldsymbol{\sigma} = 2G\sqrt{\frac{3}{2}}\left(\varepsilon'_{\alpha} - \sqrt{\frac{3}{2}}\xi_{\alpha}\right) = \sqrt{\frac{3}{2}}\sigma'_{\alpha}. \tag{5.56}$$

In order to complete the set of basic equations, the von Mises equivalent stress is given by

$$\mathcal{A}_{\text{eq}} = \sqrt{\frac{3}{2}}\|\boldsymbol{\sigma}'\|_2 = \sqrt{\frac{3}{2}}\|\hat{\sigma}'\|_2 = \|\hat{\tau}\|_2. \tag{5.57}$$

5.4.3 The uncoupled model

In the uncoupled model, each inelastic mode has an individual equivalent stress

$$\bar{\mathcal{A}}_{\text{eq},\alpha} = |\tau_{\alpha}| \le \|\hat{\tau}\|_2 = \mathcal{A}_{\text{eq}} = \sqrt{\frac{3}{2}}\|\boldsymbol{\sigma}'\|_2. \tag{5.58}$$

Incorporating a fundamental norm inequality shows that the actual equivalent stress $\mathcal{A}_{\text{eq}}$ found in the material is underestimated by its approximation $\bar{\mathcal{A}}_{\text{eq},\alpha}$. Since all further observations depend on the specific form of the potential ϕ^*, attention is limited to the rate independent case. Then each inelastic mode has an individual yield criterion

$$\varphi_{\alpha,\text{NTFA}}(\tau_{\alpha}, r(q_{\alpha})) = |\tau_{\alpha}| - (\sigma_{\text{F0}} + r(q_{\alpha})) \le 0. \tag{5.59}$$

Considering a point on the yield surface of the material at the initial hardening state $q_1, \dots, q_5 = 0$, it is possible to satisfy between one and five of the proposed yield criteria. The von Mises equivalent stress $\mathcal{A}_{\text{eq}}$ of the corresponding stress state is found in the prohibitively broad range

$$\sigma_{\text{F0}} + r(0) \le \mathcal{A}_{\text{eq}} \le \sqrt{5}\,(\sigma_{\text{F0}} + r(0)). \tag{5.60}$$

While the lower bound provides the exact solution, the von Mises stress can be overestimated by a factor of up to $\sqrt{5}$. The resulting stress is thus up 120% too high. A generalization of the error bound can be made if a n-dimensional spatial ($n = 2, 3$) setting with d-dimensional deviatoric subspace of symmetric second order tensors is examined:

$$\sigma_{\mathrm{F0}} + r(0) \le \mathcal{A}_{\mathrm{eq}} \le \sqrt{d}\,(\sigma_{\mathrm{F0}} + r(0)). \tag{5.61}$$

For three-dimensional problems $d = 5$ holds, for two-dimensional problems $d = 2$ is satisfied. This norm inequality is well-known. It relates the l_2 and l_∞ norm of a vector $\boldsymbol{x} \in \mathbb{R}^d$ via

$$\|\boldsymbol{x}\|_\infty \le \|\boldsymbol{x}\|_2 \le \sqrt{d}\|\boldsymbol{x}\|_\infty. \tag{5.62}$$

In Michel and Suquet (2003) two inelastic modes are considered for one of the example problems. The corresponding results are shown in Fig. 4 in the publication. In this two-dimensional case, the error bound is narrower than the prediction made by (5.60) due to $d = 2$. However, the initial (macroscopic) yield stress of the uncoupled model and the *exact* (in the numerical sense) solution are related by a factor of approximately $\sqrt{2} \approx 1.4$. This is the upper bound found from the theoretical observation given by (5.61). Since both, theoretical and computational evidence pointing out the upper bound character of the uncoupled model are found, the model should be disregarded from further investigations. Further, it is noteworthy that the relative error found for a homogeneous reference cell provides a good prediction of the inaccuracy of a real heterogeneous structure.

5.4.4 The coupled model

As was already stated by Michel and Suquet (2003, 2004) and confirmed, e.g., by Fritzen and Böhlke (2010b), the coupled model can capture the effective macroscopic behavior to a considerable extent. Straight-forward application of (5.57) shows*

$$\varphi_{\mathrm{NTFA}}(\hat{\tau}, \bar{r}(\bar{q})) = \|\hat{\tau}\|_2 - (\sigma_{\mathrm{F0}} + \bar{r}(\bar{q})) = \mathcal{A}_{\mathrm{eq}} - (\sigma_{\mathrm{F0}} + r(\bar{q})) = \varphi(\mathcal{A}, \bar{r}(\bar{q})). \tag{5.63}$$

Hence, the coupled model can *exactly* replicate the yield criterion of the homogeneous material, if the variables q and $\bar{q}$ coincide. Now the inelastic strain rate of the local material and of the coupled model are analyzed

$$\dot{\boldsymbol{\varepsilon}}^{\mathrm{p}} = \dot{\lambda}\sqrt{\frac{3}{2}}\frac{\boldsymbol{\sigma}'}{\|\boldsymbol{\sigma}'\|_2} = \dot{\lambda}\sqrt{\frac{3}{2}}\sum_{\alpha=1}^{5}\frac{\tau_\alpha}{\|\hat{\tau}\|_2}\boldsymbol{B}^{(\alpha)}, \qquad \dot{q} = \dot{\lambda}, \tag{5.64}$$

$$\dot{\boldsymbol{\varepsilon}}^{\mathrm{p}}_{\mathrm{NTFA}} = \dot{\lambda}_{\mathrm{NTFA}}\sum_{\alpha=1}^{5}\frac{\tau_\alpha}{\|\hat{\tau}\|_2}\boldsymbol{\mu}^{(\alpha)} = \dot{\lambda}_{\mathrm{NTFA}}\sqrt{\frac{3}{2}}\sum_{\alpha=1}^{5}\frac{\tau_\alpha}{\|\hat{\tau}\|_2}\boldsymbol{B}^{(\alpha)}, \quad \dot{q}_{\mathrm{NTFA}} = \dot{\lambda}_{\mathrm{NTFA}}. \tag{5.65}$$

In addition to this the relation

$$\dot{\lambda} = \dot{\lambda}_{\mathrm{NTFA}} \tag{5.66}$$

*The superscript index $^{(m)}$ denoting the material is omitted since only one material is present.

holds for the two Lagrangian multipliers. The evolution of both, the plastic strain tensor and the hardening variable q is, thus, identical in the NTFA material and the local model. Hence, the NTFA with the coupled model can *exactly* model the constitutive behavior of a homogeneous reference material. Further, the previous investigations together with the good numerical agreement found by Michel and Suquet (2003, 2004); Roussette et al. (2009); Fritzen and Böhlke (2010b, 2011b) motivate the use of this model for heterogeneous materials, i.e., non-uniform fields $\boldsymbol{\mu}^{(\alpha)}$. As was observed for the uncoupled model, the result obtained for the homogeneous material transfers surprisingly well to heterogeneous structures, see for instance section 6.5.2 (see also Fritzen and Böhlke, 2011b), where highly anisotropic particles have successfully been used with the coupled model.

5.4.5 A modified coupled model

The coupled model was found to give good results for a variety of problems Michel and Suquet (2003, 2004); Michel and Suquet (2009); Roussette et al. (2009); Fritzen and Böhlke (2009a,b, 2010b, 2011b). Fritzen and Böhlke (2010b, 2011b) used a minor modification of the model which concerns the normalization of the modes. For a J_2 elasto-(visco-)plastic material the mode requirement M2 was replaced by

[M2b] The modes are normalized with respect to the domain Ω_{m_α} in which they have their support

$$\langle \|\boldsymbol{\mu}^{(\alpha)}\|_2 \rangle_{\Omega_{m_\alpha}} = \frac{1}{c_{m_\alpha}} \langle \|\boldsymbol{\mu}^{(\alpha)}\|_2 \rangle = 1 \quad \Rightarrow \quad \langle \|\boldsymbol{\mu}^{(\alpha)}\|_2 \rangle = c_{m_\alpha}. \tag{5.67}$$

Assuming homogeneous modes $\boldsymbol{\mu}^{(\alpha)}$ this criterion replicates exactly the evolution of the hardening variable q. The requirement derives from

$$\bar{r}^{(m)} = \frac{\partial \bar{w}}{\partial \bar{q}^{(m)}} = c^{(m)} \varrho^{(m)} \frac{\partial \psi_{\mathrm{h}}(\bar{q}^{(m)}, \bar{\theta})}{\partial \bar{q}^{(m)}} = c^{(m)} r(\bar{q}^{(m)}, \bar{\theta}). \tag{5.68}$$

Consequently the yield criterion has to be modified to

$$\begin{aligned} \bar{\varphi}^{(m)}(\hat{\tau}^{(m)}, \bar{r}^{(m)}) &= \|\hat{\tau}^{(m)}\|_2 - \sqrt{\frac{2}{3}} (c^{(m)} \sigma_{\mathrm{F0}}^{(m)} + \bar{r}^{(m)}(\bar{q}^{(m)}, \theta)) \\ &= \|\hat{\tau}^{(m)}\|_2 - \sqrt{\frac{2}{3}} c^{(m)} (\sigma_{\mathrm{F0}}^{(m)} + r^{(m)}(\bar{q}^{(m)}, \theta)) \leq 0. \end{aligned} \tag{5.69}$$

For rate dependent materials with the rate potential $\phi^{*,(m)}(\mathcal{A}_{\mathrm{eq}}^{(m)}, r_{\mathrm{eq}}^{(m)})$ the equivalent replacement in the NTFA model is

$$\bar{\mathcal{A}}_{\mathrm{eq}}^{(m)} = \sqrt{\frac{3}{2}} \frac{\|\hat{\tau}^{(m)}\|_2}{c^{(m)}}, \qquad \bar{r}_{\mathrm{eq}}^{(m)} = \frac{\bar{r}^{(m)}}{c^{(m)}} = r^{(m)}(\bar{q}^{(m)}, \bar{\theta}), \tag{5.70}$$

$$\bar{\phi}^{*,(m)}(\hat{\tau}^{(m)}, \bar{r}^{(m)}) = c^{(m)} \phi^{*,(m)}(\bar{\mathcal{A}}_{\mathrm{eq}}^{(m)}, \bar{r}_{\mathrm{eq}}^{(m)}). \tag{5.71}$$

The partial derivatives of these equivalent replacement variables are

$$\frac{\partial \bar{A}_{\text{eq}}^{(m)}}{\partial \hat{\tau}^{(m)}} = \sqrt{\frac{3}{2}} \frac{1}{c^{(m)}} \frac{\hat{\tau}^{(m)}}{\|\hat{\tau}^{(m)}\|_2}, \qquad \frac{\partial \bar{r}_{\text{eq}}^{(m)}}{\partial \bar{r}^{(m)}} = \frac{1}{c^{(m)}}. \tag{5.72}$$

Applying the chain rule the partial derivatives of $\bar{\phi}^{*,(m)}$ are

$$\frac{\partial \bar{\phi}^{*,(m)}}{\partial \hat{\tau}^{(m)}} = \frac{\partial \bar{\phi}^{*,(m)}(\bar{A}_{\text{eq}}^{(m)}, r_{\text{eq}}^{(m)})}{\partial \bar{A}_{\text{eq}}^{(m)}} \frac{\partial \bar{A}_{\text{eq}}^{(m)}}{\partial \hat{\tau}^{(m)}} = \frac{\partial \phi^{*,(m)}(\bar{A}_{\text{eq}}^{(m)}, r_{\text{eq}}^{(m)})}{\partial \bar{A}_{\text{eq}}^{(m)}} \frac{\hat{\tau}^{(m)}}{\|\hat{\tau}^{(m)}\|_2}, \tag{5.73}$$

$$\frac{\partial \bar{\phi}^{*,(m)}}{\partial \bar{r}^{(m)}} = \frac{\partial \phi^{*,(m)}(\bar{A}_{\text{eq}}^{(m)}, r_{\text{eq}}^{(m)})}{\partial r_{\text{eq}}^{(m)}}. \tag{5.74}$$

The difference to the original criterion proposed by Michel and Suquet (2003, 2004) is subtle. In fact, the only difference is found in the evolution of the hardening variables $\bar{q}^{(m)}$. A simple example outlines the difference. Assert a single inelastic phase (so $^{(m)}$ is omitted for brevity) with phase concentration 0.5 and a single, homogeneous inelastic mode $\boldsymbol{\mu}$. Then in the original formulation $\xi = \pm\bar{q}$ holds for proportional processes. For given $\bar{q}$, e.g. for $\bar{q} = 1$, the plastic strain for the model with the original proposition M2 satisfies

$$\|\boldsymbol{\varepsilon}^{\text{p}}\|_2 = \|\boldsymbol{\mu}(\boldsymbol{x})\|_2 = \begin{cases} 0 & \text{in } \Omega_{\text{tot}} \setminus \Omega_m, \\ \sqrt{6} & \text{in } \Omega_{\text{tot}} \setminus \Omega_m. \end{cases} \tag{5.75}$$

Consequently the equivalent plastic strain has the magnitude two within the inelastic phase, although $\bar{q} = 1$ indicates a value of one. The hardening variable is thus underestimated by a factor of two. In the new normalization condition the proper values is reproduced.

Based on these observations an additional modification of the normalization condition is proposed. While the above renormalization aims at replicating the homogeneous material response to the best possible extent, it does not replicate the material response of highly heterogeneous fields as found, e.g., in open cell foams. Therefore, the active plastic volume fraction $c_{\text{a}}^{(\alpha)}$ of the α-th mode is defined as

$$c_{\text{a}}^{(\alpha)} = \langle 1_{\|\boldsymbol{\mu}^{(\alpha)}\|_2 > \delta} \rangle \leq c^{(m_\alpha)}, \tag{5.76}$$

where $\delta > 0$ is a cut-off parameter. A versatile choice is $\delta = 10^{-3\cdots-4} \max_{\boldsymbol{x}\in\Omega}\{\|\boldsymbol{\mu}_0^{(\alpha)}\|_2\}$, with $\boldsymbol{\mu}_0^{(\alpha)}$ denoting the mode before the normalization. Accordingly, M2 is modified into

[M2c] The modes are normalized with respect to the active plastic volume

$$\langle \|\boldsymbol{\mu}^{(\alpha)}\|_2 \rangle_{\Omega_{\text{a}}} = \frac{1}{c_{\text{a}}^{(\alpha)}} \langle \|\boldsymbol{\mu}^{(\alpha)}\|_2 \rangle = 1 \quad \Rightarrow \quad \langle \|\boldsymbol{\mu}^{(\alpha)}\|_2 \rangle = c_{\text{a}}^{(\alpha)}. \tag{5.77}$$

5.5 Mode identification

To this point the actual inelastic modes and their identification have not been described. More precisely all equations have assumed a given set of N modes distributed over the M constituents of the material. In the following aspects of the mode identification procedure and their implications on the constitutive equations are discussed.

For a von Mises type elasto-plastic or elasto-visco-plastic material the requirements imposed on the modes are recalled:

[H1] The support of the modes is restricted to individual constituents.

and the more specific mode restrictions

[M1] The modes are L^2 orthogonal, i.e., $\langle \boldsymbol{\mu}^{(\alpha)} \cdot \boldsymbol{\mu}^{(\beta)} \rangle = 0 \; (\alpha \neq \beta)$.

[M2c] The modes are normalized (see section 5.4.5)

$$\langle \|\boldsymbol{\mu}^{(\alpha)}\|_2 \rangle = c_{\mathrm{a}} = \langle 1_{\|\boldsymbol{\mu}^{(\alpha)}\|_2 > \delta} \rangle. \tag{5.78}$$

Based on these assumptions and their modifications, different mode identification strategies are compared in the following.

5.5.1 Numerical Experiments

The basis of all mode identification strategies are numerical experiments performed on the discretized microstructure. The boundary conditions for these experiments are not specified to this point. In Fritzen and Böhlke (2010b) the five orthonormal loadings

$$\bar{\boldsymbol{\varepsilon}}^{(\alpha)} = \varepsilon_0 \boldsymbol{B}'^{(\alpha)} \tag{5.79}$$

are suggested, where ε_0 is a reasonable load amplitude (i.e. in the range of the application of the material) and the tensors $\boldsymbol{B}'^{(\alpha)}$ denote an orthonormal basis of the five-dimensional space of deviatoric second-order tensors in $\mathbb{R}^3$. These five load steps can be completed by a purely spherical loading

$$\bar{\boldsymbol{\varepsilon}}^{(6)} = \varepsilon_1 \boldsymbol{B}^{\circ}, \tag{5.80}$$

with $\varepsilon_1 \ll \varepsilon_0$ for void-free materials and a carefully chosen $\varepsilon_1 > 0$ in the case of voided materials.

A consideration of stress-driven or mixed loading conditions was, in the considered cases, not necessary to replicate the macroscopic stress response with good quality. However, some new load cases of practical importance are, for example

[L1] Uni-axial stress conditions $\bar{\boldsymbol{\sigma}}(t) = t\dot{\sigma}_0 \, \boldsymbol{n} \otimes \boldsymbol{n}$ resembling a tension test in direction $\boldsymbol{n}$;

[L2] Simple shear $\bar{\boldsymbol{\sigma}}(t) = t\dot{\bar{\sigma}}_0 \,\mathrm{sym}(\boldsymbol{m} \otimes \boldsymbol{n}), \boldsymbol{n} \cdot \boldsymbol{m} = 0$ resembling a torsion test;

[L3] Bi-axial loading with varying stress ratio;

The first two of these load cases can easily be realized in a strain-driven setting when only the homogeneous part of the displacements in the loading direction are prescribed (case L1). An analogous procedure is possible for L2.

While a large number of test cases could be considered in general, the computational cost of the virtual experiments is usually non-negligible. Restrictions in terms of CPU time and memory apply in most cases. Supposing that a basis $\{\boldsymbol{\mu}^{(\alpha)}\}$ $(\alpha = 1, \ldots, N)$ has already been identified a new virtual experiment is performed. Let the plastic strain field in this calculation (for simplicity a single time frame is considered) be denoted $\boldsymbol{\varepsilon}^{\mathrm{p}}$. In order to judge on the efficiency of the basis $\boldsymbol{\mu}^{(\alpha)}$ with respect to the new field data, the value

$$\delta_\alpha = \frac{\langle \boldsymbol{\mu}^{(\alpha)} \cdot \boldsymbol{\varepsilon}^{\mathrm{p}} \rangle}{\langle \boldsymbol{\mu}^{(\alpha)} \cdot \boldsymbol{\mu}^{(\alpha)} \rangle} \tag{5.81}$$

can be calculated for $\alpha = 1, \ldots, N$, i.e. the projection of the plastic strain on the subspace of all possible plastic deformations is computed. The associated residual inelastic strain is

$$\boldsymbol{r} = \boldsymbol{\varepsilon}^{\mathrm{p}} - \sum_{\alpha=1}^{N} \delta_\alpha \boldsymbol{\mu}^{(\alpha)}. \tag{5.82}$$

If the condition

$$\langle \|\boldsymbol{r}\|_2 \rangle \leq \delta_{\mathrm{p}} \tag{5.83}$$

is satisfied, then the field $\boldsymbol{\varepsilon}^{\mathrm{p}}$ can be considered negligible for a suitably chosen δ_{p}, e.g., $\delta_{\mathrm{p}} \approx 10^{-2} \langle \|\boldsymbol{\varepsilon}^{\mathrm{p}}\|_2 \rangle$. Hence, the efficiency of a determined basis can easily be verified for additional virtual experiments *a posteriori*.

5.5.2 Karhunen-Loève decomposition

In the context of the NTFA the Karhunen-Loève decomposition was first proposed by Roussette et al. (2009). It can, however, be considered a standard procedure used in order reduction methods in the context of the POD (proper orthogonal decomposition, see also section 5.2). The following description assumes the existence of N_{s} different numerical data sets (equivalent to preprocessing calculations). Each of these contains snapshots of the entire plastic strain field $\boldsymbol{\varepsilon}^{\mathrm{p},(\alpha)}(\boldsymbol{x}, t_\beta)$ $(\alpha = 1, \ldots, N_{\mathrm{s}})$ at each discrete time point t_β, where β can be considered as the field frame number and $1 \leq \beta \leq N_{\mathrm{f}}^{(\alpha)}$ with $N_{\mathrm{f}}^{(\alpha)}$ the total number of frames in set α. In order to simplify the notation the total number of recorded fields

$$N_{\mathrm{f}} = \sum_{\alpha=1}^{N_{\mathrm{s}}} N_{\mathrm{f}}^{(\alpha)} \tag{5.84}$$

is introduced. Then two index functions $I(\alpha)$, $J(\alpha)$ can be defined, such that

$$\boldsymbol{\varepsilon}^{\mathrm{p}}_{\alpha}(\boldsymbol{x}) = \boldsymbol{\varepsilon}^{\mathrm{p},(I(\alpha))}(\boldsymbol{x}, t_{J(\alpha)}) \quad (\alpha = 1, \ldots, N_{\mathrm{f}}) \tag{5.85}$$

is a one parametric mapping associating with each generalized index α a single snapshot of the plastic strain from the given data basis. In the NTFA the Karhunen-Loève decomposition is then based on a proper value decomposition of the matrix

$$(\hat{K}_{\mathrm{p}})_{\alpha\beta} = \langle \boldsymbol{\varepsilon}^{\mathrm{p}}_{\alpha} \cdot \boldsymbol{\varepsilon}^{\mathrm{p}}_{\beta} \rangle \quad (\alpha, \beta = 1, \ldots N_{\mathrm{f}}) \tag{5.86}$$

of the plastic strain fields observed at all time points simulated in the numerical experiments. Note that $\langle \boldsymbol{A} \cdot \boldsymbol{B} \rangle$ has the same meaning as the Euclidean L^2 inner product for tensors on the domain Ω_{p} occupied by the plastic phase. The matrix $\hat{K}_{\mathrm{p}}$ is symmetric and positive semi-definite by definition. Hence, a set of real-valued non-negative eigenvalues v_α and corresponding eigenvectors $\hat{v}^{(\alpha)}$ exists ($\alpha = 1, \ldots, N_{\mathrm{f}}$). It is found that the eigenvalues v_α decrease quickly, see, e.g., Roussette et al. (2009) or Fig. 5.3.

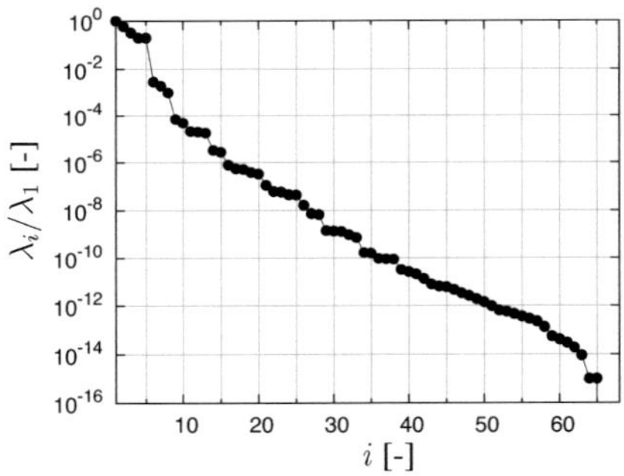

Figure 5.3: Normalized eigenvalues of the matrix $\hat{K}_{\mathrm{p}}$ of the plastic strain fields of the microstructures B.O4 (section 6.5)

The eigenvectors associated with the N largest eigenvalues can be used to construct a $N-$dimensional basis of inelastic modes according to

$$\boldsymbol{\mu}_0^{(\alpha)} = \sum_{\beta=1}^{N_{\mathrm{f}}} (\hat{v}^{(\alpha)})_{\beta} \boldsymbol{\varepsilon}^{\mathrm{p}}_{\beta} \quad (\alpha = 1, \ldots, N), \tag{5.87}$$

and a subsequent renormalization. For $\alpha, \beta = 1, \ldots, N_{\mathrm{f}}$ the equality

$$\langle \boldsymbol{\mu}_0^{(\alpha)} \cdot \boldsymbol{\mu}_0^{(\beta)} \rangle = \hat{v}^{(\alpha)} \cdot \hat{K}_{\mathrm{p}} \hat{v}^{(\beta)} = \left\{ \begin{array}{ll} 0 & \alpha \neq \beta, \\ \lambda_\alpha & \alpha = \beta. \end{array} \right. \tag{5.88}$$

holds. The created modes are, hence, orthogonal (hypothesis M1). A brief summary on the algorithm is presented in Fig. 5.4.

[KL1] Compute the components of the matrix $\hat{K}_\mathrm{p}$

$$(\hat{K}_\mathrm{p})_{ij} = (\hat{K}_\mathrm{p})_{ji} = \langle \boldsymbol{\varepsilon}_i^\mathrm{p} \cdot \boldsymbol{\varepsilon}_j^\mathrm{p} \rangle \quad (i = 1, \ldots, N_\mathrm{f},\ j = i, \ldots, N_\mathrm{f}). \tag{5.89}$$

[KL2] Compute the eigenvalues λ_i $(i = 1, \ldots, N_\mathrm{f})$ of $\hat{K}_\mathrm{p}$ and let $\lambda_i \geq \lambda_j$ for $j > i$.

[KL3] Find for a given cut-off parameter $1 > \delta_\mathrm{p} > 0$ the largest N, such that

$$\lambda_N > \delta_\mathrm{p} \lambda_1 \tag{5.90}$$

is satisfied for all $j = 1, \ldots, N$ the criterion.

[KL4] Compute the orthogonal eigenvectors $\hat{v}^{(\alpha)}$ to the eigenvalues λ_α.

[KL5] Construct the modes by linear combination of the plastic strain fields

$$\boldsymbol{\mu}_0^{(\alpha)} = \sum_{i=1}^{N_\mathrm{f}} (\hat{v}^{(\alpha)})_i \boldsymbol{\varepsilon}_i^\mathrm{p}. \tag{5.91}$$

[KL6] Renormalize the modes with respect to the active plastic volume (M2c)

$$\boldsymbol{\mu}^{(\alpha)} = \frac{c_\mathrm{a}^\alpha}{\langle \| \boldsymbol{\mu}_0^{(\alpha)} \|_2 \rangle_{\Omega_{\alpha,\mathrm{a}}}} \boldsymbol{\mu}_0^{(\alpha)}. \tag{5.92}$$

Figure 5.4: Algorithm for the Karhunen-Loève based kinematic mode identification approach

5.5.3 Alternative kinematic mode identification procedure

Fritzen and Böhlke (2010b) proposed a modified mode identification procedure that is based on a Gram-Schmidt procedure. The decision whether or not an existing basis is going to be enlarged based on a new plastic strain field $\boldsymbol{\varepsilon}^\mathrm{p}$ is taken by computation of the residual obtained from the projection algorithm described in section 5.5.1. It was found by Fritzen and Böhlke (2010b) that cycling through all computations in reverse order of the computed frames, i.e. starting with the last computed field, the modes could quickly be identified. Moreover, an enrichment of the basis using the proposed residual criterion is always possible at small computational cost. Particularly, the matrix $\hat{K}_\mathrm{p}$ does not need to be recomputed which results in large computational savings. In the Karhunen-Loève based approach, the addition of new modes requires not only the computation of new entries in the matrix, but also the recomputation of all inelastic modes, since the eigenvectors of $\hat{K}_\mathrm{p}$ change. Thereby the solution of all associated eigenfields etc. is mandatory. The algorithm is presented in Fig. 5.5 and described in detail in Fritzen and Böhlke (2010b).

[MI1] Initialize $N = 0$. Assert that the N_f computed fields are arranged as follows:

- First strain path calculation; Last frame to first frame
- Second strain path calculation; Last frame to first frame
- ...

Set $i = 0$ and choose tolerance $\delta_0 > 0$.

[MI2] Increment i and compute

$$r_0 = \langle \boldsymbol{\varepsilon}_i^\text{P} \cdot \boldsymbol{\varepsilon}_i^\text{P} \rangle. \tag{5.93}$$

Cycle through $\alpha = 1, \ldots, N$ and compute

$$k_\alpha = \langle \boldsymbol{\mu}_0^{(\alpha)} \cdot \boldsymbol{\varepsilon}_i^\text{P} \rangle, \tag{5.94}$$

as long as

$$r_\alpha = r_{\alpha-1} - (k_\alpha)^2 \tag{5.95}$$

exceeds a prescribed offset δ_0. If $r_\alpha < \delta_0$ go to MI2.

[MI3] If $r_N > \delta$ then add a new mode:

$$\boldsymbol{\mu}^{(N+1)} = \frac{1}{\sqrt{r_N}} \left(\boldsymbol{\varepsilon}_i^\text{P} - \sum_{\alpha=1}^{N} k_\alpha \boldsymbol{\mu}_0^{(\alpha)} \right). \tag{5.96}$$

Increment N and go to MI2.

[MI4] Renormalize the modes with respect to the active plastic volume (M2c)

$$\boldsymbol{\mu}^{(\alpha)} = \frac{c_\text{a}^\alpha}{\langle \| \boldsymbol{\mu}_0^{(\alpha)} \|_2 \rangle_{\Omega_{\alpha,\text{a}}}} \boldsymbol{\mu}_0^{(\alpha)}. \tag{5.97}$$

Figure 5.5: Algorithm for the alternative kinematic mode identification algorithm

5.5.4 A thermodynamically motivated approach

Based on the observations of the properties of the matrix $\hat{D}(\bar{\theta})$, i.e. its negative-definiteness, a modified mode identification procedure can be proposed. In the proposed ansatz it is assumed that the kinematic hardening is temperature independent, i.e. $\hat{K}(\bar{\theta}) \equiv \hat{K}(\bar{\theta}_0)$. Then the matrix $\hat{D}$ is constant, i.e. it is not depending on the macroscopic temperature $\bar{\theta}$. The approach does not refer to the kinematic restriction M1 but does replace it by

[M1b] The modes with support in the same constituent are orthogonal with respect to the weighted L^2-inner product (for a given reference temperature $\bar{\theta}_0$)

$$[\boldsymbol{\mu}^{(\alpha)}, \boldsymbol{\mu}^{(\beta)}]_p = -\langle \boldsymbol{\sigma}_*^{(\alpha)} \cdot \mathbb{S}[\boldsymbol{\sigma}_*^{(\beta)}] \rangle - k(\bar{\theta}_0) \langle \boldsymbol{\mu}^{(\alpha)} \cdot \mathbb{K}[\boldsymbol{\mu}^{(\beta)}] \rangle. \tag{5.98}$$

The modified normalization condition M2c is not altered in the presented approach. Note that $\boldsymbol{\sigma}_*^{(\alpha)}$ is a linear function of $\boldsymbol{\mu}^{(\alpha)}$. Hence the bi-linear properties of the operator $[\bullet, \bullet]_p$ are guaranteed. Due to the symmetry and negative definiteness of $\hat{D}(\bar{\theta}_0)$ (see section 5.3) the operator $[\bullet, \bullet]_p$ in (5.98) always denotes an inner product.

The proposed approach is due to a thermodynamical observation: If the entry $D_{\alpha\beta}$ of the mode interaction matrix is zero for $\alpha, \beta \in \mathcal{I}_m$ $(m = 1, \ldots, M)$, this implies an independence of the inelastic mechanisms inducing the respective inelastic strain fields. Interestingly the approach does not include any kinematic restrictions but does solely depend on the structure of the thermodynamic driving forces $\hat{\tau}$ (see section 5.1, (5.25)).

A numerical implication of the preceding observation is the particular structure of the matrix $\hat{D}$ (Fig. 5.6). The block matrices representing the interaction of modes within one constituent are diagonal matrices. In the case of a single inelastic constituent the outcome is a fully diagonal matrix $\hat{D}$. As a consequence the efficiency of the algorithms presented

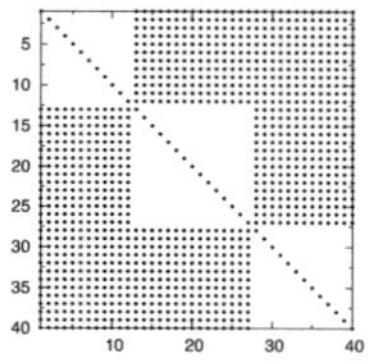

Figure 5.6: Structure of the matrix $\hat{D}$ for the thermodynamically motivated mode identification procedure (3 Materials, 40 modes in total, $\bullet$ denotes a non-zero entry)

in section 5.6.3 can be significantly improved by exploiting the block diagonal structure and the obtained numerical performance observed for the homogenized material behavior of heterogeneous materials can be considerably improved. It is noteworthy that the entries $D_{\alpha\beta}$ for $\alpha \in \mathcal{I}_{m_1}$, $\beta \in \mathcal{I}_{m_2}$ are usually non-zero. This implies that interactions between the inelastic mechanisms found within different constituents do generally exist.

The entries describing the interaction between phases ($\alpha, \beta = 1, \ldots, N$ and $m_\alpha \neq m_\beta$) are often rather small in amplitude. In order to show the latter an upper bound of the entries can be derived based on

$$\begin{aligned} D_{\alpha\beta} &= \left|\langle \boldsymbol{\mu}^{(\alpha)} \cdot \mathbb{C}[\boldsymbol{\varepsilon}_*^{(\beta)} - \boldsymbol{\mu}^{(\beta)}]\rangle - k(\bar{\theta}_0)\langle \boldsymbol{\mu}^{(\alpha)} \cdot \mathbb{K}[\boldsymbol{\mu}^{(\beta)}]\rangle\right| \\ &= \left|\langle \boldsymbol{\mu}^{(\alpha)} \cdot \mathbb{C}[\boldsymbol{\varepsilon}_*^{(\beta)}]\rangle - \langle \boldsymbol{\mu}^{(\alpha)} \cdot (\mathbb{C} + k(\bar{\theta}_0)\mathbb{K})[\boldsymbol{\mu}^{(\beta)}]\rangle\right| \\ &= \left|\langle \boldsymbol{\mu}^{(\alpha)} \cdot \mathbb{C}[\boldsymbol{\varepsilon}_*^{(\beta)}]\rangle\right| = c_{m_\alpha} \left|\langle \boldsymbol{\mu}^{(\alpha)} \cdot \mathbb{C}[\boldsymbol{\varepsilon}_*^{(\beta)}]\rangle_{\Omega_{m_\alpha}}\rangle\right| \\ &\leq c_{m_\alpha} 2G_{m_\alpha} \langle \|\boldsymbol{\mu}^{(\alpha)}\|_2\rangle_{\Omega_{m_\alpha}} \langle \|\boldsymbol{\varepsilon}_*^{(\beta)}\|_2\rangle_{\Omega_{m_\alpha}} \leq 2G_{m_\alpha} \langle \|\boldsymbol{\varepsilon}_*^{(\beta)}\|_2\rangle_{\Omega_{m_\alpha}}. \end{aligned} \tag{5.99}$$

Due to having their support in different constituents, either $\boldsymbol{\mu}^{(\alpha)}$ or $\boldsymbol{\mu}^{(\beta)}$ are zero at all points $\boldsymbol{x} \in \Omega$. Considering additionally the symmetry of $\hat{D}$, an improved estimate is

$$D_{\alpha\beta} \leq \min_{\gamma,\delta \in \{\alpha,\beta\};\gamma\neq\delta} \left\{ 2G_{m_\delta} \langle \|\boldsymbol{\varepsilon}_*^{(\gamma)}\|_2\rangle_{\Omega_{m_\delta}} \right\}. \tag{5.100}$$

Similar observations help to understand why the modes obtained from the presented process are almost orthogonal with respect to the standard inner product L^2_Ω. This can be shown by considering for $\alpha, \beta \in \mathcal{I}_m$

$$\begin{aligned} D_{\alpha\beta} = 0 &= \langle \boldsymbol{\mu}^{(\alpha)} \cdot \mathbb{C}[\boldsymbol{\varepsilon}_*^{(\beta)} - \boldsymbol{\mu}^{(\beta)}]\rangle - k(\bar{\theta}_0)\langle \boldsymbol{\mu}^{(\alpha)} \cdot \mathbb{K}[\boldsymbol{\mu}^{(\beta)}]\rangle \\ &= \langle \boldsymbol{\mu}^{(\alpha)} \cdot \mathbb{C}[\boldsymbol{\varepsilon}_*^{(\beta)}]\rangle - \langle \boldsymbol{\mu}^{(\alpha)} \cdot (\mathbb{C} + k(\bar{\theta}_0)\mathbb{K})[\boldsymbol{\mu}^{(\beta)}]\rangle \\ \Rightarrow \ \langle \boldsymbol{\mu}^{(\alpha)} \cdot \boldsymbol{\mu}^{(\beta)}\rangle &= \frac{2G_m}{(2G_m + k(\bar{\theta}_0)K_m)} \langle \boldsymbol{\mu}^{(\alpha)} \cdot \boldsymbol{\varepsilon}_*^{(\beta)}\rangle \\ &\leq c_m \frac{2G_m}{(2G_m + k(\bar{\theta}_0)K_m)} \langle \|\boldsymbol{\mu}^{(\alpha)}\|_2\rangle_{\Omega_m} \langle \|\boldsymbol{\varepsilon}_*^{(\beta)}\|_2\rangle_{\Omega_m} \\ &\leq \frac{2G_m}{(2G_m + k(\bar{\theta}_0)K_m)} \langle \|\boldsymbol{\varepsilon}_*^{(\beta)}\|_2\rangle_{\Omega_m} \leq \langle \|\boldsymbol{\varepsilon}_*^{(\beta)}\|_2\rangle_{\Omega_m}. \end{aligned} \tag{5.101}$$

Switching α and β a stricter inequality can be obtained

$$\langle \boldsymbol{\mu}^{(\alpha)} \cdot \boldsymbol{\mu}^{(\beta)}\rangle \leq \frac{2G_m}{(2G_m + k(\bar{\theta}_0)K_m)} \min_{\gamma=\alpha,\beta} \langle \|\boldsymbol{\varepsilon}_*^{(\gamma)}\|_2\rangle_{\Omega_m} \leq \min_{\gamma=\alpha,\beta} \langle \|\boldsymbol{\varepsilon}_*^{(\gamma)}\|_2\rangle_{\Omega_m}. \tag{5.102}$$

Accordingly the modes can be considered near L^2_Ω-orthogonal in the sense of the proposition M1 of Michel and Suquet (2003). Notably the latter is exactly satisfied for the homogeneous material. It can thus easily be verified that the minimum requirement proposed in section 5.4.2 is satisfied. More precisely the inelastic behavior of a homogeneous body is *exactly* replicated by the NTFA model with the modified orthogonality condition M1b.

The implementation of this identification strategy is involved. Here, a method involving the Karhunen-Loève proper orthogonal decomposition is proposed. More precisely, a staggered mode identification procedure is considered. First, the kinematic approach is pursued to find a set of N inelastic modes which are L^2_Ω orthogonal. Second, the eigensolutions $\boldsymbol{\sigma}_*, \boldsymbol{\varepsilon}_*$ are computed. Third, the symmetric matrix $\hat{D}(\bar{\theta}_0)$ is computed. Based on a proper value decomposition of the symmetric and positive matrix $\hat{B} = -\hat{D}(\bar{\theta}_0)$ a set of eigenmodes that are orthonormal with respect to $[\bullet, \bullet]_p$ is constructed. Details are given in Fig. 5.7.

[MI1] Compute a set of N modes $\boldsymbol{\mu}_0^{(\alpha)}$ used the Karhunen-Loève based approach (see Fig. 5.4).

[MI2] Solve for the fields $\boldsymbol{\sigma}_{*,0}^{(\alpha)}, \boldsymbol{\varepsilon}_{*,0}^{(\alpha)}$ induced by the modes.

[MI3] Compute the matrix $\hat{D}(\bar{\theta}_0)$, set $\hat{B} = -\hat{D}(\bar{\theta}_0)$ and solve the eigenproblem

$$(\hat{B} - \lambda_\alpha \hat{I})\hat{v}^{(\alpha)} = 0. \tag{5.103}$$

[MI4] Compute the new modes

$$\boldsymbol{\mu}_1^{(\alpha)} = \sum_{\beta=1}^{N} (\hat{v}^{(\alpha)})_\beta \boldsymbol{\mu}_0^{(\beta)}. \tag{5.104}$$

Perform an analogous linear combination for the associated fields $\boldsymbol{\sigma}_*, \boldsymbol{\varepsilon}_*, \boldsymbol{u}_*$.

[MI5] Renormalize the modes with respect to the active plastic volume (M2c)

$$\boldsymbol{\mu}^{(\alpha)} = \frac{c_{\rm a}^\alpha}{\langle \|\boldsymbol{\mu}_1^{(\alpha)}\|_2 \rangle_{\Omega_{\alpha,\rm a}}} \boldsymbol{\mu}_1^{(\alpha)}. \tag{5.105}$$

Figure 5.7: Thermodynamically motivated mode identification algorithm based on a staggered POD scheme

5.6 Numerical implementation of the non-uniform transformation field analysis

5.6.1 Data processing steps

The actual determination of the homogenized material response based on the NTFA consists of a series of individual steps. While the properties of the method have been analyzed in the previous sections in detail, the actual implementation has not been addressed. All details on the implementation are provided in the following. An overview on the individual steps is given in Fig. 5.8.

The first step consists in the modeling of the material. This stage includes the discretization of the geometry and the definition of the local elasto-(visco-)plastic material behavior. Homogeneous specimens made of the individual components can be used to identify the individual characteristics in experiments, e.g. the elastic moduli and the hardening parameters.

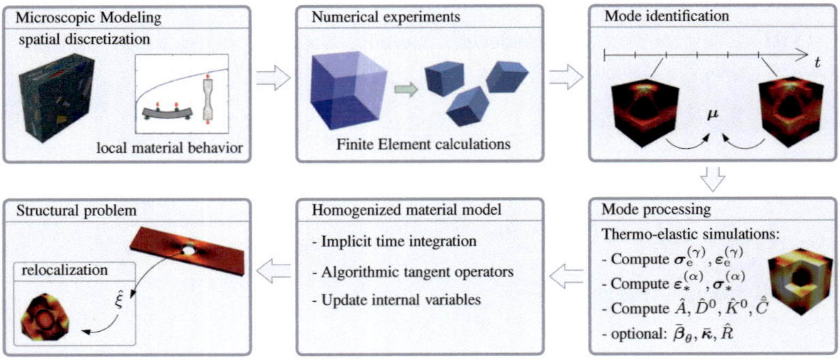

Figure 5.8: Overview on all steps of the NTFA

In the second step numerical experiments are performed on the microheterogeneous material defined in the first step. Detailed information on the underlying procedure is provided in section 5.5.1. These simulations can take a substantial amount of time and the number of different loadings should be chosen appropriately. In particular the kinematic or static load amplitude that is expected to act on the material in real world conditions should be considered. The resulting output may consist of considerable amounts of data, i.e., many gigabytes of disk space are required for a single microstructure (numbers are provided, e.g., in Fritzen and Böhlke (2011b)).

Based on the evidence obtained in the virtual testing the mode identification algorithm (section 5.5) is started. The aim is the identification of a possibly small number of characteristic plastic deformation fields. The finite element operators needed here and thereafter are described in section 5.6.2. The output of this stage are the plastic strain fields $\boldsymbol{\mu}^{(\alpha)}$. The input data from the virtual experiments is no longer needed and could be deleted at this point. Thereby the amount of stored data can be reduced to an acceptable level.

The plastic strain fields are not sufficient to provide all matrices that are characteristic for the homogenized material. In fact, a set of N eigenstress problems has to be solved in order to get the fields $\boldsymbol{\sigma}_*^{(\alpha)}$ and $\boldsymbol{\varepsilon}_*^{(\alpha)}$. Additionally, six calculations with prescribed strains are required to provide the linear elastic properties. In a thermo-mechanically coupled analysis the required fields $\boldsymbol{\sigma}_\theta$ and the thermal conductivity need also be computed. At the end the following parameters are the only quantities required for the homogenized material model:

- Thermo-elastic properties: The effective elastic stiffness tensor $\bar{\mathbb{C}}$ ($\equiv \hat{\hat{C}}$), the thermo-elastic stress $\bar{\boldsymbol{\beta}}_\theta$ ($\equiv \hat{\hat{\beta}}_\theta$) and the effective heat conductivity tensor $\bar{\boldsymbol{\kappa}}$ ($\equiv \hat{\hat{\kappa}}$);
- Inelastic material parameters of the homogenized model: the matrices $\hat{A}$, $\hat{D}^0$, $\hat{K}^0$, $\hat{R}$;
- Phase concentrations $c^{(\alpha)}$.

All of these parameters are completely independent of a data fitting procedure or other inverse problems. Indeed, they are all derived based on the rigorous micro-mechanical observations presented earlier. In particular the topology and the material symmetry is decoupled from the identification process.

The resulting homogenized material model can be used for different purposes. One application is found in terms of an individual test suite to investigate the behavior of the heterogeneous material in an isolated environment. This allows for example to draw the yield surface of the macroscopic material (see section 6.5.2). Another major application is found in structural computations based on the finite element method. Here the homogenized material model can be applied at each integration point associated with the actual microheterogeneous material. Note that the macroscopic structure may also contain other materials (e.g., section 6.5.4) or different types of NTFA materials. Thereby it can be accounted for a spatially heterogeneous distribution of the microstructure as found, e.g., in functionally graded materials. Equivalently, the model can be used with the FFT (fast fourier transformation) method or FDM (finite difference method).

An interesting feature at this stage is the possibility to perform a relocalization of the stress, the total and the inelastic strain and the displacement field for any vector $\hat{\xi}(t)$. This reconstruction is based on the localization operators presented in (5.7)-(5.9). It allows for the investigation of stress localizations in the material or for a statistical investigation of the stress distribution in individual constituents at selected points of a macroscopic structure. The thereby obtained information can be used to predict failure of the material.

In order to use the NTFA material in the presented cases, a robust and efficient numerical implementation of the time discrete constitutive equations of the model is described in the sections 5.6.3-5.6.4.

5.6.2 Finite element operators

In order to compute the components of the system matrices $\hat{A}, \hat{D}^0, \hat{K}^0$ and in order to perform the mode identification procedure, a variety of different operations need to be performed on the fields. These operations involve (possibly weighted) L^2_Ω-products, linear combinations and volume averages of fields or functions of fields. When the NTFA is used with a standard finite element code then these operators are usually unavailable. Definitions of all operators are provided in the following which allow for an implementation into most existing finite element codes.

Linear combination. The field $\Theta(\boldsymbol{x})$ computed from a linear combination of m individual field frames $\Theta^{(i)}(\boldsymbol{x})$ $(i = 1, \dots, m)$ (e.g. at different time steps in an analysis) is defined by

$$\Theta(\boldsymbol{x}) = \sum_{i=1}^{m} k_i \Theta^{(i)}(\boldsymbol{x}), \tag{5.106}$$

with real-valued coefficients $k_i \in \mathbb{R}$. In discrete form, the combination of fields can be computed by node-wise or integration point-wise summation of the individual components of the field variable. More specifically, displacements are defined on the nodes, whereas the stress and strain tensor and the inelastic variables are associated with integration points. For example, if the field $\Theta(\boldsymbol{x})$ is scalar valued with value $\Theta_j = \Theta(\boldsymbol{x}_j)$ at the j-th out of n_p nodes (or integration points) then the discrete form of the linear combination is defined as

$$\Theta_j = \sum_{i=1}^{m} k_i \Theta_j^{(i)} \qquad (j = 1, \ldots, n_\mathrm{p}). \tag{5.107}$$

Vectors, matrices and tensors are treated analogously with the only modification being that the summation has to be carried out for each component of a vector or matrix. Additionally, it is necessary to define a common basis when computing the sum of tensor components.

Averaging operator. The discrete equivalent of the averaging operator $\langle\bullet\rangle$ can be defined for data found at integration points. Therefore, it is needed to introduce for the integration point $\boldsymbol{x}_j$ with index j its associated integration volume λ_j. Then the volume average of a scalar field $\Theta(\boldsymbol{x})$ on a domain $P \subseteq \Omega$ is approximated by

$$\langle\Theta\rangle_P = \frac{1}{|P|} \int_P \Theta(\boldsymbol{x})\,\mathrm{d}V \approx \frac{1}{V(P)} \left(\sum_{i \in I(P)} \Theta_i \lambda_i \right). \tag{5.108}$$

Here $I(P)$ is a set containing the indices of all integration points contained in P, and $V(P) = \sum_{i \in I(P)} \lambda_i$ is the volume of P as computed by the approximated integration procedure.

L^2_P-product. The L^2_P product ($P \subseteq \Omega$) of two vector fields $\boldsymbol{v}(\boldsymbol{x}), \boldsymbol{w}(\boldsymbol{x})$ is defined as

$$\langle \boldsymbol{v} \cdot \boldsymbol{w} \rangle_P = \frac{1}{|P|} \int_P \boldsymbol{v}(\boldsymbol{x}) \cdot \boldsymbol{w}(\boldsymbol{x})\,\mathrm{d}V. \tag{5.109}$$

Substituting the exact volume average by the approximated operator (5.108) with the scalar field $\Theta = \boldsymbol{v} \cdot \boldsymbol{w}$ an algorithm for the numerical computation of $\langle \boldsymbol{v} \cdot \boldsymbol{w} \rangle_P$ is

$$\langle \boldsymbol{v} \cdot \boldsymbol{w} \rangle_P \approx \frac{1}{V(P)} \left(\sum_{i \in I(P)} \boldsymbol{v}(\boldsymbol{x}_i) \cdot \boldsymbol{w}(\boldsymbol{x}_i) \lambda_i \right). \tag{5.110}$$

Correspondingly the L^2_P product for tensor fields $\boldsymbol{A}, \boldsymbol{B}$ is

$$\langle \boldsymbol{A} \cdot \boldsymbol{B} \rangle_P \approx \frac{1}{V(P)} \left(\sum_{i \in I(P)} \boldsymbol{A}(\boldsymbol{x}_i) \cdot \boldsymbol{B}(\boldsymbol{x}_i) \lambda_i \right). \tag{5.111}$$

Weighted averages are carried out by replacing the standard inner product. An example of the latter is the product $[\bullet, \bullet]_p$ defined in (5.98).

5.6.3 Implicit time integration for rate-independent materials

In the following the implementation of the homogenized material response of a microstructured material consisting of M inelastic constituents with rate-independent inelastic behavior is presented. Attention is limited to the coupled model and the modified mode normalization criterion M2c (section 5.4.5). It is asserted that for each of the constituents a number of N_i modes has been identified. The total number of inelastic modes is

$$N = \sum_{i=1}^{M} N_i. \tag{5.112}$$

For convenience the numbering of the modes is assumed ordered by constituent. In particular the modes of the first constituent are the modes numbered $\{1, \ldots, N_1\}$. The second material then has the (global) mode indices $\{N_1 + 1, \ldots, N_1 + N_2\}$ and so on. The focus is on GSM (see section 2.2.3) with one scalar hardening variable $\bar{q}^{(m)}$ for the m-th material. The resulting evolution equations for the internal variables are

$$\begin{aligned} \dot{\hat{\xi}}^{(m)} &= \dot{\lambda}_m \frac{\partial \bar{\varphi}^{(m)}(\hat{\tau}^{(m)}, \bar{r}^{(m)})}{\partial \hat{\tau}^{(m)}}, & \dot{\bar{q}}^{(m)} &= \dot{\lambda}_m \frac{\partial \bar{\varphi}^{(m)}(\hat{\tau}^{(m)}, \bar{r}^{(m)})}{\partial \bar{r}^{(m)}}, \\ \dot{\lambda}_m &\geq 0 & \bar{\varphi}^{(m)}(\hat{\tau}^{(m)}, \bar{q}^{(m)}) &\leq 0, \end{aligned} \tag{5.113}$$

where $\dot{\lambda}_m$ $(m = 1, \ldots, M)$ are Lagrangian multipliers enforcing the admissibility of the thermodynamic state. The reduced vector $\hat{\tau}^{(m)}$ contains all components of $\hat{\tau}$ associated with the m-th micro-material and $\bar{q}^{(m)}$ is the corresponding hardening variable.

The trial vector of the conjugate forces resulting from a converged state $(\hat{\tau}(t_n), \boldsymbol{\varepsilon}(t_n), \hat{\xi}(t_n), \hat{q}(t_n), \bar{\theta}(t_n))$ subjected to the load increment $(\Delta\bar{\varepsilon}, \Delta\bar{\theta})$ when assuming purely elastic response is

$$\hat{\tau}_{\text{tr}} = \hat{\tau}(t_n) + \hat{A}\Delta\hat{\bar{\varepsilon}} + \hat{R}\Delta\bar{\theta}. \tag{5.114}$$

In order to abbreviate the notation considerably, the following quantities are introduced:

$$\hat{n}_m = \frac{\partial \bar{\varphi}^{(m)}}{\partial \hat{\tau}^{(m)}}, \qquad \hat{N} = \begin{pmatrix} \hat{n}_1 & \hat{0} & \ldots & \\ \hat{0} & \hat{n}_2 & \hat{0} & \ldots \\ \vdots & \vdots & \vdots & \vdots \end{pmatrix}, \tag{5.115}$$

$$\hat{\Lambda} = \Delta t \begin{pmatrix} \dot{\lambda}_1 \\ \vdots \\ \dot{\lambda}_M \end{pmatrix} = \begin{pmatrix} \Delta\lambda_1 \\ \vdots \\ \Delta\lambda_2 \end{pmatrix}, \qquad \Delta\bar{q}^{(m)} = \sqrt{\frac{2}{3}}\Delta\lambda_m. \tag{5.116}$$

Applying the implicit backward Euler time integration results in the root finding problem

$$\hat{f}(\hat{\tau}, \hat{\Lambda}) = \begin{pmatrix} \hat{\tau} - \hat{D}\hat{N}\hat{\Lambda} - \hat{\tau}_{\text{tr}} \\ -\bar{\varphi}^{(1)}(\hat{\tau}^{(1)}, \bar{q}^{(1)}(t_n) + \Delta\bar{q}^{(1)}) \\ \vdots \\ -\bar{\varphi}^{(M)}(\hat{\tau}^{(M)}, \bar{q}^{(M)}(t_n) + \Delta\bar{q}^{(M)}) \end{pmatrix} = \hat{0}. \tag{5.117}$$

The Jacobian $\dfrac{\partial \hat{f}}{\partial(\hat{\tau}, \hat{\Lambda})}$ of $\hat{f}$ is given by

$$\hat{J} = \left(\begin{array}{c|cccc} \hat{I} - \hat{D}\,\text{BlDiag}\left(\dfrac{\partial^2 \bar{\varphi}^{(m)}}{\partial \hat{\tau}^{(m)} \partial \hat{\tau}^{(m)}}\right)\hat{\Lambda} & & -\hat{D}\hat{N} & \\ \hline & -\sqrt{\dfrac{2}{3}}\dfrac{\partial \bar{\varphi}^{(1)}}{\partial \bar{q}^{(1)}} & 0 & \cdots \\ -\hat{N}^{\mathsf{T}} & 0 & -\sqrt{\dfrac{2}{3}}\dfrac{\partial \bar{\varphi}^{(2)}}{\partial \bar{q}^{(2)}} & 0 \\ & \vdots & 0 & \ddots \end{array} \right) \tag{5.118}$$

Introducing the additional abbreviations

$$\hat{H} = \begin{pmatrix} -\sqrt{\dfrac{2}{3}}\dfrac{\partial \bar{\varphi}^{(1)}}{\partial \bar{q}^{(1)}} & 0 & \cdots \\ 0 & -\sqrt{\dfrac{2}{3}}\dfrac{\partial \bar{\varphi}^{(2)}}{\partial \bar{q}^{(2)}} & 0 \\ \vdots & 0 & \ddots \end{pmatrix}, \tag{5.119}$$

$$\hat{X} = \begin{pmatrix} \Delta\lambda_1 \dfrac{\partial^2 \bar{\varphi}^{(1)}}{\partial \hat{\tau}^{(1)} \partial \hat{\tau}^{(1)}} & & \\ & \Delta\lambda_2 \dfrac{\partial^2 \bar{\varphi}^{(2)}}{\partial \hat{\tau}^{(2)} \partial \hat{\tau}^{(2)}} & \\ & & \ddots \end{pmatrix}, \tag{5.120}$$

$$\frac{\partial^2 \bar{\varphi}^{(m)}}{\partial \hat{\tau}^{(m)}\, \partial \hat{\tau}^{(m)}} = \frac{1}{\|\hat{\tau}^{(m)}\|_2} \left(\hat{I}_{N_m} - \frac{\hat{\tau}^{(m)} (\hat{\tau}^{(m)})^{\mathsf{T}}}{\hat{\tau}^{(m)} \cdot \hat{\tau}^{(m)}} \right), \tag{5.121}$$

the matrix $\hat{J}$ can be expressed in terms of the multiplicative decomposition

$$\hat{J}_* = \begin{pmatrix} \hat{D}^{-1} - \hat{X} & -\hat{N} \\ -\hat{N}^{\mathsf{T}} & \hat{H} \end{pmatrix}, \qquad \hat{J} = \begin{pmatrix} \hat{D} & \\ & \hat{I}_M \end{pmatrix} \hat{J}_*. \tag{5.122}$$

Since the matrix $\hat{D}$ is negative definite (see also section 5.3), its inverse can always be computed. Moreover, $\hat{D}^{-1}$ can be determined to virtually arbitrary precision in a preprocessing step and then be reused in each increment*. The symmetry of $\hat{J}_*$ is a direct result of the presented straight-forward computation and the Schwartz theorem. Based on the Jacobian and its decomposition into two symmetric matrices in (5.122), a Newton scheme can be used to efficiently compute the vector $\hat{\tau}(t_{n+1})$ and the incremental Lagrangian multipliers $\hat{\Lambda}$. During the Newton iteration, the values of $\bar{\varphi}^{(m)}$ have to be set to zero, if the m-th material is elastic in this increment. The complete scheme is shown in Fig. 5.9.

*This is only true in the case where $\hat{D}$ is independent of the temperature. If not so, then $\hat{D}$ is unconditionally constant during the local time integration and its inverse can be computed once and stored throughout the local iteration process.

[S1] Compute $\hat{\tau}_{\text{tr}}$ based on load increment $\Delta\bar{\varepsilon}, \Delta\bar{\theta}$.
Update macroscopic stress according to $\Delta\bar{\boldsymbol{\sigma}} = \bar{\mathbb{C}}[\Delta\bar{\varepsilon}] + \bar{\boldsymbol{\beta}}_\theta \Delta\bar{\theta}$.

[S2] Evaluate $\bar{\varphi}^{(m)}(\hat{\tau}_{\text{tr}}^{(m)}, \bar{r}^{(m)}(t_n))$.
If all micro materials are elastic ($\bar{\varphi}^{(m)} \leq 0$): set $\mathbb{C}_{\text{a}} = \bar{\mathbb{C}}$ and exit.

[S3] Initialize, $\hat{\Lambda} = 0, \hat{\tau} = \hat{\tau}_{\text{tr}}$.

[S4] Compute residual vector $\hat{f}$, with $f_{m+N} = 0$, if $\Delta\lambda_m = 0$ and $\bar{\varphi}^{(m)} \leq 0$.
Recompute $\hat{J}_*$.

[S5] Check convergence $\|\hat{f}\|_R < \delta_R$, with δ_R a convergence parameter and

$$\|\hat{f}\|_R = \sum_{m=1}^{M} \sqrt{\sum_{\alpha \in \mathcal{I}_m} (f_\alpha)^2} + \|\hat{\bar{\varphi}}\|_1. \tag{5.123}$$

If converged goto **[S7]**.

[S6] Solve the linear system

$$\hat{f} + \hat{J} \begin{pmatrix} \Delta\hat{\tau} \\ \Delta\hat{\Lambda} \end{pmatrix} = 0, \tag{5.124}$$

and update the vector $\hat{\tau} \rightarrow \hat{\tau} + \Delta\hat{\tau}$ and the Lagrangian multipliers $\hat{\Lambda} \rightarrow \hat{\Lambda} + \Delta\hat{\Lambda}$. Goto **[S4]**.

[S7] Update the internal variables $\hat{\xi}_{n+1} = \hat{\xi}_n + \hat{N}\hat{\Lambda}$, $\bar{q}_{n+1}^{(m)} = \bar{q}_n^{(m)} + \sqrt{2/3}\,\Delta\lambda_m$.
Update the macroscopic stress tensor $\bar{\boldsymbol{\sigma}}$ based on (5.126).

[S8] Compute algorithmic tangent operator (section 5.6.4) and exit.

Figure 5.9: Newton scheme for the implicit time integration of the homogenized material behavior

The obtained solution provides the increment in the mode activity coefficients by

$$\Delta\hat{\xi} = \hat{N}\hat{\Lambda}^{\mathsf{T}} = \begin{pmatrix} \Delta\lambda_1 \hat{n}_1 \\ \Delta\lambda_2 \hat{n}_2 \\ \vdots \end{pmatrix}. \tag{5.125}$$

Thereby, the macroscopic stress update is computed to

$$\bar{\boldsymbol{\sigma}}(t_{n+1}) = \bar{\boldsymbol{\sigma}}(t_n) + \bar{\mathbb{C}}[\Delta\bar{\varepsilon}] + \sum_{\alpha=1}^{N} \Delta\xi_\alpha \bar{\boldsymbol{\sigma}}_*^{(\alpha)} + \bar{\boldsymbol{\beta}}_\theta \Delta\bar{\theta}. \tag{5.126}$$

Using (5.40) the computationally handy matrix-vector notation

$$\hat{\bar{\sigma}}(t_{n+1}) = \hat{\bar{\sigma}}(t_n) + \hat{\bar{C}}\Delta\hat{\bar{\varepsilon}} - \hat{A}^{\mathsf{T}}\Delta\hat{\xi} + \hat{\bar{\beta}}_\theta\Delta\bar{\theta} \tag{5.127}$$

is obtained.

5.6.4 Algorithmic tangent operators for rate-independent materials

The algorithmic tangent operators with respect to strain and temperature increments are valuable quantities. In the following computations the dependency $\Delta\bar{\boldsymbol{\sigma}} \equiv \Delta\bar{\boldsymbol{\sigma}}(\bar{\varepsilon},\bar{\theta})$ is exploited, i.e., the increments of the internal variables depend only on the previous state and the increment in the macroscopic kinematic and thermal loading. Then the differential form of (5.126) is

$$\begin{aligned} \mathrm{d}\bar{\boldsymbol{\sigma}}(t_{n+1}) &= \bar{\mathbb{C}}[\,\mathrm{d}\boldsymbol{\varepsilon}] + \sum_{\alpha=1}^{N} \bar{\boldsymbol{\sigma}}_*^{(\alpha)} \otimes \frac{\partial\Delta\xi_\alpha}{\partial\bar{\boldsymbol{\varepsilon}}}\,\mathrm{d}\bar{\boldsymbol{\varepsilon}} + \sum_{\alpha=1}^{N} \bar{\boldsymbol{\sigma}}_*^{(\alpha)}\,\frac{\partial\Delta\xi_\alpha}{\partial\bar{\theta}}\,\mathrm{d}\bar{\theta} + \bar{\boldsymbol{\beta}}_\theta\,\mathrm{d}\bar{\theta} \\ &= \underbrace{\left(\bar{\mathbb{C}} + \sum_{\alpha=1}^{N} \bar{\boldsymbol{\sigma}}_*^{(\alpha)} \otimes \frac{\partial\Delta\xi_\alpha}{\partial\bar{\boldsymbol{\varepsilon}}}\right)}_{=\bar{\mathbb{C}}_\mathrm{a}}\mathrm{d}\bar{\boldsymbol{\varepsilon}} + \underbrace{\left(\sum_{\alpha=1}^{N} \bar{\boldsymbol{\sigma}}_*^{(\alpha)}\,\frac{\partial\Delta\xi_\alpha}{\partial\bar{\theta}} + \bar{\boldsymbol{\beta}}_\theta\right)}_{=\bar{\boldsymbol{\beta}}_\mathrm{a}}\mathrm{d}\bar{\theta}. \end{aligned} \tag{5.128}$$

From a theoretical background it is known, that generalized standard materials result in a symmetric tangent operator with respect to strain changes (see, e.g., Hackl, 1996). In order to prove this symmetry of the tangent operator explicitly for the considered materials and in the proposed implicit time integration scheme, a series of mathematical operations have to be considered. Therefore, the general form of the tangent operator is derived in the first place. Accounting for the previous calculation, one can compute

$$\left.\frac{\partial\hat{f}}{\partial\Delta\hat{\bar{\varepsilon}}}\right|_{\hat{\tau},\hat{\Lambda},\bar{\theta}} = \begin{pmatrix} -\hat{A} \\ \hat{0} \end{pmatrix}, \qquad \left.\frac{\partial\hat{f}}{\partial\Delta\bar{\theta}}\right|_{\hat{\tau},\hat{\Lambda},\bar{\varepsilon}} = \begin{pmatrix} -\hat{R} \\ -\dfrac{\partial\hat{\bar{\varphi}}}{\partial\bar{\theta}} \end{pmatrix}. \tag{5.129}$$

The derivative of the yield criteria with respect to temperature is given by

$$\frac{\partial\bar{\varphi}^{(m)}}{\partial\bar{\theta}} = -\sqrt{\frac{2}{3}}\,c^{(m)}\,\frac{\partial r^{(m)}(\bar{q}^{(m)},\bar{\theta})}{\partial\bar{\theta}}. \tag{5.130}$$

By the chain rule the identity

$$\begin{pmatrix} \hat{A} & \hat{R} \\ \hat{0} & \dfrac{\partial\hat{\bar{\varphi}}}{\partial\bar{\theta}} \end{pmatrix} = \begin{pmatrix} \hat{D} & \\ & \hat{I}_M \end{pmatrix}\hat{J}_* \begin{pmatrix} \dfrac{\partial\hat{\tau}}{\partial\hat{\bar{\xi}}} & \dfrac{\partial\hat{\tau}}{\partial\bar{\underline{\theta}}} \\ \dfrac{\partial\hat{\Lambda}}{\partial\hat{\bar{\varepsilon}}} & \dfrac{\partial\hat{\Lambda}}{\partial\bar{\theta}} \end{pmatrix}, \tag{5.131}$$

is found. Based on

$$\Delta\hat{\xi} = \hat{N}\hat{\Lambda}, \qquad \left(\frac{\partial \Delta\hat{\xi}}{\partial \hat{\tau}} \quad \frac{\partial \Delta\hat{\xi}}{\partial \hat{\Lambda}} \right) = \left(\hat{X} \quad \hat{N} \right). \tag{5.132}$$

the derivatives with respect to strain and temperature changes are

$$\begin{aligned} \left(\frac{\partial \Delta\hat{\xi}}{\partial \hat{\bar{\varepsilon}}} \quad \frac{\partial \Delta\hat{\xi}}{\partial \bar{\theta}} \right) &= \left(\hat{X}, \hat{N} \right) \begin{pmatrix} \frac{\partial \hat{\tau}}{\partial \hat{\bar{\xi}}} & \frac{\partial \hat{\tau}}{\partial \bar{\theta}} \\ \frac{\partial \hat{\Lambda}}{\partial \hat{\bar{\varepsilon}}} & \frac{\partial \hat{\Lambda}}{\partial \bar{\theta}} \end{pmatrix} \\ &= \left(\hat{X}, \hat{N} \right) \hat{J}_*^{-1} \begin{pmatrix} \hat{D}^{-1}\hat{A} & \hat{D}^{-1}\hat{R} \\ \hat{0} & \frac{\partial \hat{\bar{\varphi}}}{\partial \bar{\theta}} \end{pmatrix}, \end{aligned} \tag{5.133}$$

Using the property (5.40) of the matrix $\hat{A}$ (see section 5.3) the algorithmic tangent operators are explicitly given by

$$\bar{\mathbb{C}}_{\mathrm{a}} \rightarrow \hat{\bar{C}}_{\mathrm{a}} = \hat{\bar{C}} - \hat{A}^{\mathsf{T}} \frac{\partial \Delta\hat{\xi}}{\partial \hat{\bar{\varepsilon}}}, \tag{5.134}$$

$$\bar{\boldsymbol{\beta}}_{\mathrm{a}} \rightarrow \hat{\beta}_{\mathrm{a}} = \hat{\bar{\beta}}_{\theta} - \hat{A}^{\mathsf{T}} \frac{\partial \Delta\hat{\xi}}{\partial \bar{\theta}}. \tag{5.135}$$

Symmetry of $\hat{\bar{C}}_{\mathrm{a}}$. Due to the symmetry of $\bar{\mathbb{C}}$ (and hence $\hat{\bar{C}}$) it is sufficient to show the symmetry of

$$\Delta\hat{C} = \hat{\bar{C}}_{\mathrm{a}} - \hat{\bar{C}} = -\hat{A}^{\mathsf{T}} \frac{\partial \Delta\hat{\xi}}{\partial \hat{\bar{\varepsilon}}} = -\hat{A}^{\mathsf{T}} \left(\hat{X}, \hat{N} \right) \hat{J}_*^{-1} \begin{pmatrix} \hat{D}^{-1}\hat{A} \\ \hat{0} \end{pmatrix}. \tag{5.136}$$

Since the second part $\hat{J}_*$ is symmetric, it's inverse is also symmetric and can be written as

$$\hat{J}_*^{-1} = \begin{pmatrix} \hat{U} & \hat{V} \\ \hat{V}^{\mathsf{T}} & \hat{W} \end{pmatrix}, \tag{5.137}$$

with symmetric matrices $\hat{U} \in Sym(\mathbb{R}^N), \hat{W} \in Sym(\mathbb{R}^M)$, and a rectangular matrix $\hat{V} \in \mathbb{R}^{N \times M}$. In a first step, the matrix product

$$\hat{J}_*^{-1}\hat{J}_* = \begin{pmatrix} \hat{U}\hat{D}^{-1} - \hat{U}\hat{X} - \hat{V}\hat{N}^{\mathsf{T}} & -\hat{U}\hat{N} + \hat{V}\hat{H} \\ -\hat{N}^{\mathsf{T}}\hat{U} + \hat{H}\hat{V}^{\mathsf{T}} & \hat{V}^{\mathsf{T}}\hat{V} + \hat{W}\hat{H} \end{pmatrix} = \hat{I}_{N+M} \tag{5.138}$$

is evaluated. It follows by block comparison of the first $N \times N$ components, that

$$\hat{U}\hat{X} + \hat{V}\hat{N}^{\mathsf{T}} = -\hat{I}_N + \hat{U}\hat{D}^{-1} \tag{5.139}$$

holds. Secondly, (5.136) is rewritten as

$$
\begin{aligned}
-\hat{A}^{\mathsf{T}} \frac{\partial \Delta \hat{\xi}}{\partial \hat{\bar{\varepsilon}}} &= -\hat{A}^{\mathsf{T}} \left(\hat{X}, \hat{N} \right) \hat{J}_{*}^{-1} \begin{pmatrix} \hat{D}^{-1} \hat{A} \\ \hat{0} \end{pmatrix} \\
&= -\hat{A}^{\mathsf{T}} \left(\left(\hat{X}, \hat{N} \right) \hat{J}_{*}^{-1} \begin{pmatrix} \hat{D}^{-1} \\ \hat{0} \end{pmatrix} \right) \hat{A}.
\end{aligned} \tag{5.140}
$$

Due to the symmetric application of $\hat{A}$, it suffices to show the symmetry of the inner bracket. Substitution of (5.137) into this matrix equation yields the identity

$$
\left(\hat{X}, \hat{N} \right) \begin{pmatrix} \hat{U} \hat{D}^{-1} \\ \hat{V}^{\mathsf{T}} \hat{D}^{-1} \end{pmatrix} = \left(\hat{X} \hat{U} + \hat{N} \hat{V}^{\mathsf{T}} \right) \hat{D}^{-1}. \tag{5.141}
$$

Resorting to the transpose of (5.139), the identity

$$
\left(\hat{X} \hat{U} + \hat{N} \hat{V}^{\mathsf{T}} \right) \hat{D}^{-1} = \left(-\hat{I}_N + \hat{U} \hat{D}^{-1} \right)^{\mathsf{T}} \hat{D}^{-1} \quad = -\hat{D}^{-1} + \hat{D}^{-1} \hat{U} \hat{D}^{-1}. \tag{5.142}
$$

is established. Due to the symmetry of $\hat{D}$ and $\hat{U}$, the last expression is symmetric and, thereby, the tangent stiffness operator is *always* symmetric for the considered class of materials.

Chapter 6

Computational results

6.1 Homogenization of crystalline aggregates

6.1.1 Elastic properties of cubic crystal aggregates

The mesh generation technique presented in section 3.3.1.3 has been used in extensive numerical studies on polycrystals by Fritzen et al. (2009). Periodic ensembles containing 10, 20, 50 and 100 grains are examined with respect to their effective elastic behavior. Example discretizations are shown in Fig. 6.1.

Figure 6.1: Example discretizations used in the numerical study

The crystallographic orientation is assumed constant throughout individual grains and cubic elastic symmetry is considered. The stiffness tensor of each grain in such an aggregate can be denoted by (see, e.g., Rychlewski, 1995)

$$\mathbb{C} = 3K\mathbb{P}_1 + \mu_1\mathbb{P}_2 + \mu_2\mathbb{P}_3, \quad \mathbb{P}_1 = \frac{1}{3}\boldsymbol{I} \otimes \boldsymbol{I}, \quad \mathbb{P}_2 = \mathbb{D} - \mathbb{P}_1, \quad \mathbb{P}_3 = \mathbb{I}^{s} - \mathbb{D}. \quad (6.1)$$

Here K is the isotropic bulk modulus of the crystal and μ_1, μ_2 are two distinct eigenvalues associated with deviatoric terms. The tensor $\mathbb{D}$ is a structural tensor defined by the orthonormal lattice vectors $\boldsymbol{g}_i$ ($i = 1, 2, 3$) of the single crystal via

$$\mathbb{D} = \sum_{i=1}^{3} \boldsymbol{g}_i \otimes \boldsymbol{g}_i \otimes \boldsymbol{g}_i \otimes \boldsymbol{g}_i. \tag{6.2}$$

A uniform distribution on the group of rotations $SO(3)$ is used to assign random orientations to the grains of the aggregate. The latter is realized based on random variables X, Y, Z that are uniformly distributed on the interval $[0, 1)$ to determine three Euler angles (z-x-z convention) via (see for instance Murnaghan, 1962)

$$\varphi_1 = 2\pi X, \qquad \Phi = \operatorname{acos}(2Y - 1), \qquad \varphi_2 = 2\pi Z. \tag{6.3}$$

The consideration of a uniform distribution of the crystal orientation is equivalent to an isotropic crystal orientation function (CODF). In other investigations, e.g., by Bertram and Böhlke (1999) or Böhlke and Bertram (2001), the anisotropy due to an anisotropic and possibly evolving CODF are investigated. For a cubic material the Zener ratio

$$\zeta = \frac{\mu_2}{\mu_1} \tag{6.4}$$

is an index for the anisotropy of the single crystal. In the isotropic case ζ equals one. In the presented numerical studies three different cubic materials are considered: copper, gold and nickel. The material parameters and the Zener ratio of these materials are given in Tab. 6.1.

	K [GPa]	μ_1 [GPa]	μ_2 [GPa]	ζ
Copper	136.67	47	150	3.19
Gold	167	27	79.4	2.94
Nickel	183.67	101	248	2.46

Table 6.1: Material parameters and Zener ratio of copper, gold and nickel single crystals (Simmons and Wang, 1971; Paufler and Schulze, 1978)

For each number of grains and for each of the three materials a number of 200 linear elastic homogenizations have been performed using ABAQUS/Standard in combination with a user material sub-routine (UMAT). For each of these 2'400 numerical experiments the apparent stiffness tensor $\mathbb{C}^{\mathrm{a}}$ has been determined. The apparent isotropic bulk and shear modulus are determined by means of the projection

$$3K^{\mathrm{a}} = \mathbb{C}^{\mathrm{a}} \cdot \mathbb{P}_1, \qquad 2G^{\mathrm{a}} = \frac{1}{5}\mathbb{C}^{\mathrm{a}} \cdot (\mathbb{I}^{\mathrm{s}} - \mathbb{P}_1). \tag{6.5}$$

The procedure of taking a large number of smaller unit cells is motivated by the works of Kanit (2003); Kanit et al. (2003, 2006), which showed that thereby the results of computations performed on large samples can be approximated with computationally improved efficiency. The distribution of K^{a} and G^{a} has been determined in terms of the average value $\bar{K}, \bar{G}$ and the standard deviation $\sigma(K^{\mathrm{a}}), \sigma(G^{\mathrm{a}})$ over all realizations. Since the bulk modulus of cubic materials is independent of the crystallographic orientation distribution function, the only parameter that needs to be determined is the effective shear modulus and its standard deviation. The effective Young's modulus $\bar{E}$ and the Poisson ratio $\bar{\nu}$ are frequently referred to in engineering applications. They are computed based on $\bar{K}, \bar{G}$ via

$$\bar{E} = \frac{9\bar{K}\bar{G}}{3\bar{K}+\bar{G}}, \qquad \bar{\nu} = \frac{3\bar{K}-2\bar{G}}{2(3\bar{K}+\bar{G})}. \tag{6.6}$$

The results of the simulations are compared to bounds of odd order derived by Dederichs and Zeller (1973); Zeller and Dederichs (1973). A generalization in terms of bounds of even order was published by Kröner (1977). For a comprehensive overview on the homogenization of the elastic properties of polycrystals it is referred to, e.g., Adams and Olson (1998) and the concise introduction to the topic in section 4.3. Isotropic statistical properties are assumed for the considered aggregates. The results of the numerical computations and the analytical homogenization are presented in Fig. 6.2(a), 6.2(b), 6.2(c). The fifth-order bounds have been used for comparison since seventh- and higher order bounds did not lead to notable changes.

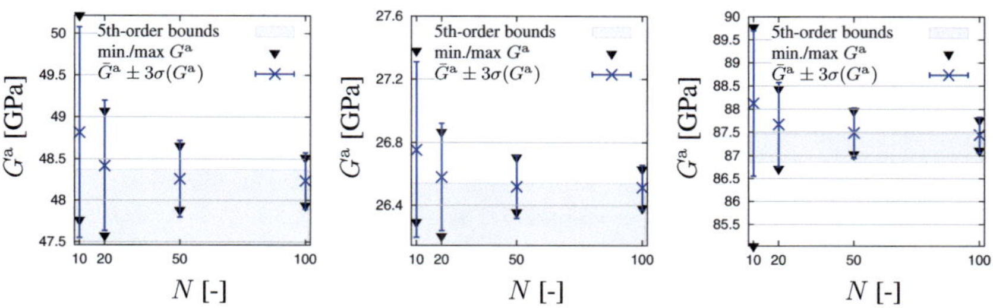

(a) copper (b) gold (c) nickel

Figure 6.2: Average value ($\times$), minimum and maximum over all realizations and $\pm 3\sigma(G^{\mathrm{a}})$ error indicators; gray area: 5-th order bounds (cf. Fritzen et al., 2009)

Based on (6.6) the Young's modulus has been determined for each realization of the polycrystal and for the effective polycrystal. Histograms of the Young's modulus based on the aggregates containing 100 grains are shown in Fig. 6.3. It is found that the Young's modulus resembles a normal distribution and that the computed standard deviation is small, i.e. in the sub-percent range of the average Young's modulus. For further details see Fritzen et al. (2009).

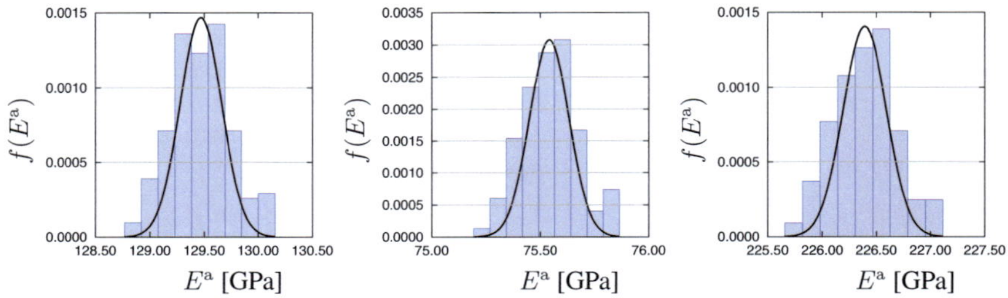

(a) copper (b) gold (c) nickel

Figure 6.3: Histogram of the Young's modulus for aggregates of 100 grains with comparison to a normal distribution with same first and second moment (cf. Fritzen et al., 2009)

6.1.2 Influence of the grain morphology on the elasto-visco-plastic properties

Based on a manipulation of the grain morphology (see section 3.3.1.2) it is possible to isolate the influence of crystallographic and morphological texture found in polycrystalline metals such as, e.g., rolled sheet metals. The morphological anisotropy for these materials is approximated by

$$\bar{\boldsymbol{F}} = \begin{pmatrix} \lambda & 0 & 0 \\ 0 & \lambda^{-1} & 0 \\ 0 & 0 & 1 \end{pmatrix} \boldsymbol{e}_i \otimes \boldsymbol{e}_j, \qquad \lambda \geq 1. \tag{6.7}$$

The influence of such a morphological texture has been investigated by Böhlke et al. (2009) for polycrystalline copper ($C_{1111} = 114$ GPa, $C_{1122} = 65.3$ GPa, $C_{1212} = 28.5$ GPa). The unit cell in the computations consisted of an ensemble containing 50 grains. Five different degrees of anisotropy ($\lambda = 1, 2^{1/4}, 2^{1/2}, 2^{3/4}, 2$) have been considered with two different micromorphologies associated with each value of λ:

[C1] the deformed grain shape,

[C2] the undeformed grain shape.

The discretization for $\lambda = 2$ is shown in Fig. 6.4. For each λ both of these microstructures are equipped with the same crystal orientation distribution function obtained from Taylor type rigid visco-plastic computations in a large strain formulation. The Taylor model is chosen for simplicity although it may over- or underestimate certain texture components. A more sophisticated model is for example presented by Böhlke (2005). The thereby generated initial medium is assumed stress free and is used in elasto-visco-plastic crystal plasticity calculations in a kinematically linear context. The evolution of the plastic slip for the considered

Figure 6.4: Example mesh for the anisotropic mesh with $\lambda = 2$ (50 grains)

face centered cubic crystal is determined by a classical Schmid law of the form

$$\dot{\boldsymbol{\varepsilon}}^{\mathrm{p}} = \sum_{\alpha=1}^{12} \dot{\gamma}_\alpha \boldsymbol{M}^{(\alpha)}, \qquad \boldsymbol{M}^{(\alpha)} = \mathrm{sym}(\boldsymbol{d}^{(\alpha)} \otimes \boldsymbol{n}^{(\alpha)}), \tag{6.8}$$

$$\tau_\alpha = \boldsymbol{\sigma} \cdot \boldsymbol{M}^{(\alpha)}, \qquad \dot{\gamma}_\alpha = \dot{\gamma}_0 \,\mathrm{sgn}(\tau_\alpha) \left\langle \frac{|\tau_\alpha| - \tau_\alpha^{\mathrm{C}}}{\tau^{\mathrm{D}}} \right\rangle^m, \quad \langle \bullet \rangle = \max\{0, \bullet\}. \tag{6.9}$$

Here τ^{D} denotes a visco-plastic overstress and τ_α^{C} is a constant critical resolved shear stress. The rate $\dot{\gamma}_0$ is a reference value for the material and $m \geq 1$ denotes the Norton exponent. In the considered example the inelastic material parameters of copper were estimated to

$$\tau^{\mathrm{D}} = 5\,\mathrm{MPa}, \qquad \tau_\alpha^{\mathrm{C}} = 40\,\mathrm{MPa}, \qquad \dot{\gamma}_0 = 10^{-3}\,\mathrm{s}^{-1}, \qquad m = 100. \tag{6.10}$$

Each of the nine different microheterogeneous materials is subjected to an isochoric kinematic loading of the type

$$\dot{\bar{\boldsymbol{\varepsilon}}} = \dot{\varepsilon}_0 \begin{pmatrix} 2 & 0 & 0 \\ 0 & -1 & 0 \\ 0 & 0 & -1 \end{pmatrix} \boldsymbol{e}_i \otimes \boldsymbol{e}_j, \qquad \dot{\varepsilon}_0 = \sqrt{\frac{2}{3}} \dot{\gamma}_0. \tag{6.11}$$

Periodic displacement fluctuation boundary conditions are imposed on the unit cells. The strain rate $\dot{\varepsilon}_0$ and the simulation time are chosen such that a macroscopic strain of $\|\bar{\varepsilon}\|_2 = 6.12\%$ is attained at the end of the simulation time. An illustration of the process is shown in Fig. 6.5. An interpretation of the virtual experiments is given in terms of taking specimens in a continuous rolling process at different levels of thickness reduction. On these specimens, tension tests are performed in order to compare the results to the ones of virtual test pieces being subjected only to a crystallographic texture, but an isotropic grain shape.
The normal components of the effective stress tensor at the end of the simulation performed on the material with and without the crystallographic texture are compared in Fig. 6.6 for all five considered values of λ.

Discussion. The numerical simulations performed on the crystal aggregates show only a small scatter of the apparent elastic properties for different random realizations of the microstructure or different crystal orientation distributions. The scatter is small, even for microstructures containing only ten or twenty grains. These observations are clearly related

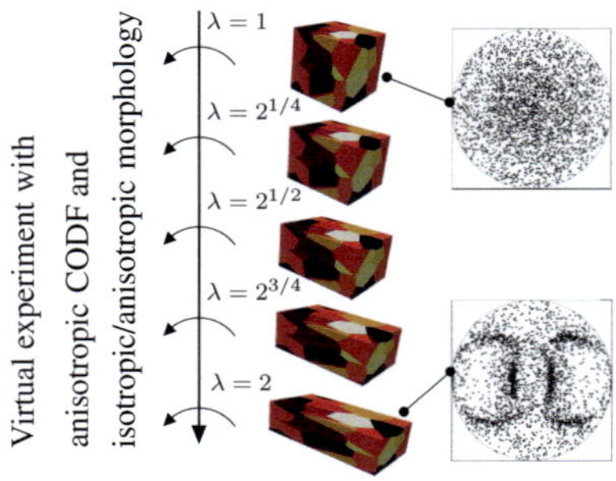

Figure 6.5: Illustration of the rolling process and the numerical experiments performed on pre-deformed specimens

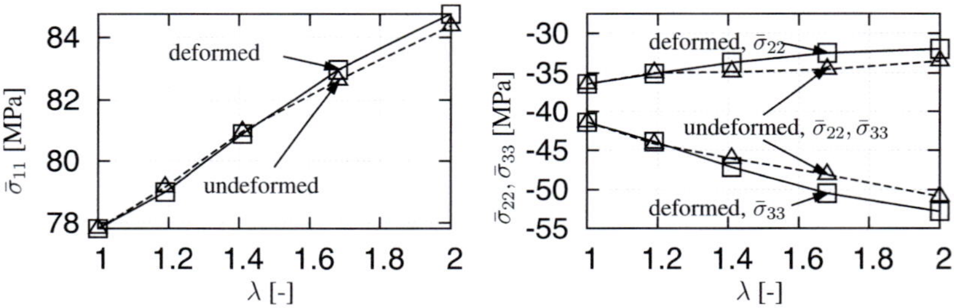

Figure 6.6: Asymptotic effective stress components in $\boldsymbol{e}_1 \otimes \boldsymbol{e}_1$-direction (left) and $\boldsymbol{e}_2 \otimes \boldsymbol{e}_2, \boldsymbol{e}_3 \otimes \boldsymbol{e}_3$ directions (right); solid line: deformed grain morphology, dashed line: undeformed grain morphology

to the periodic displacement fluctuation boundary conditions used in the simulations. The periodicity of the discretization is imposed by the mesh generation technique developed in section 3.3.1.3 (see also Fritzen et al., 2009). It allows to prescribe this type of boundary condition in an efficient way via linear relationships of master/slave degrees of freedom. An implementation of boundary conditions of this type is possible in most finite element codes, e.g. via the `*EQUATION` keyword in ABAQUS.

Based on the comparison of the computational results to the bounds of odd order of Dederichs and Zeller (1973); Zeller and Dederichs (1973), two important conclusions are possible. First, the bounds of fifth (or higher) order are sufficiently narrow for actual polycrystalline metals, i.e. for a moderate elastic anisotropy. Second, the computational prediction of the effective stiffness is found within these bounds. For isotropic microstructures the use of the bounds of fifth order is, thus, highly encouraged for materials with moderate phase contrast. However, the limitation to isotropic microstructures is a significant limitation, e.g., when materials with anisotropic CODF and/or morphology are considered. The latter has been considered in order to isolate the impact of the morphological anisotropy onto the elasto-viscoplastic material response of polycrystalline copper. It is found that the crystallographic texture influences the asymptotic stress response to a considerable extent. The changes of the grain morphology induce some changes in the effective stress response of the material. These influences are, however, small compared to the influence of the CODF.

An advantage of the presented mesh generation technique is that the resulting model can be used with arbitrary constitutive models. Observations as previously described are hardly possible with analytical models, even when introducing simplifying assumptions. For problems of this kind the proposed mesh generation technique provides a tool allowing for the consideration of realistic albeit artificial microstructures and opens up the possibility to run computational studies on statistical microstructural realizations.

6.2 Thermo-elastic properties of metal ceramic composites

Metal matrix composites (MMC) with particulate reinforcement have favorable tribological and thermo-mechanical properties (see, e.g., Suresh, 2002; Chawla and Chawla, 2006). Common material pairings for this class of composites are aluminum reinforced by silicon-carbide particles (Al/SiC) or alumina particles (Al/Al_2O_3). Possible fields of applications for these materials are found in engine and aerospace design where the beneficial effects overcome the increased manufacturing expenses. The effective mechanical properties of MMCs are, thus, of technical significance. Additionally, the understanding of the influence of the local topology on the macroscopic constitutive behavior can help to improve the efficiency of engineering structures.

Due to the underlying manufacturing process the particles in such materials are often polyhedral in shape, and an approximation in terms of the model microstructures presented in section 3.3.2 is possible (see, e.g., Suresh, 2002; Miserez et al., 2004; Chawla and Chawla, 2006). With regard to their improved resistance over standard alloys at elevated temperatures, MMCs are often used in thermo-mechanical environments. Therefore, the thermo-elastic interactions are considered in the following in addition to the effective elastic response. More precisely, the apparent stiffness tensor $\mathbb{C}^{\mathrm{a}}$ and the apparent thermal expansion coefficient $\boldsymbol{\alpha}_\theta^{\mathrm{a}}$ are evaluated for all considered microstructures. The finite element method is used to solve the seven thermo-elastic load cases described in section 4.2. Based on the expressions (4.27) and (4.40) the apparent properties of the material are obtained from the computed stress and strain fields via volume averaging over the unit cell.

The focus is on morphologically isotropic model microstructures as presented in section 3.3.2, i.e., statistically isotropic particle shapes are considered. Therefore, the effective stiffness tensor of the material is also isotropic and the effective thermal expansion coefficient has only one parameter. The computed apparent properties approximate the effective values.

In total, the three isotropic material properties

$$K^{\mathrm{a}} = \frac{1}{3}\mathbb{C}^{\mathrm{a}} \cdot \mathbb{P}_1, \qquad G^{\mathrm{a}} = \frac{1}{10}\mathbb{C}^{\mathrm{a}} \cdot (\mathbb{I}^{\mathrm{s}} - \mathbb{P}_1), \qquad \alpha_\theta^{\mathrm{a}} = \frac{1}{3}\boldsymbol{\alpha}_\theta^{\mathrm{a}} \cdot \boldsymbol{I} \tag{6.12}$$

are evaluated. A Monte Carlo type study based on ensembles containing 20 particles is performed. The periodicity of both, the geometry and the mesh are enforced using the methods presented earlier. Due to the periodicity of the unit cells combined with the sufficient number of particles within the individual cells, the different realizations are assumed to show apparent properties close to the effective ones. Following the observations by Kanit et al. (2003, 2006), the effective response of the material at a prescribed volume fraction is efficiently approximated by taking the average over all realizations containing the same particle volume fraction.

A self-written finite element implementation was used to perform the numerical simulations. Iterative linear solvers were used in terms of ILUPACK (sequential version, Bollhoefer and Saad (2006)) and HIPS (parallel version, `http://hips.gforge.inria.fr`) in order to solve the high-dimensional linear systems resulting from the finite element analysis. The efficiency of the iterative solvers was found to be excellent, particularly when considering the low memory requirements which are pronounced for periodic problems when using direct solution methods. Aspects of direct and iterative solution methods have been investigated by Fritzen and Böhlke (2010a), where different types of boundary conditions and their numerical implications are discussed.

	E [GPa]	ν [-]	α_θ [10^{-6}K^{-1}]
Al	70	0.33	25
SiC	400	0.19	4.3

Table 6.2: Material properties used in the thermo-elastic simulations for Al/SiC composites (Chawla et al., 2006, cf. to)

The material pairing considered in the following is Al/SiC. The properties of the individual constituents are taken from Chawla et al. (2006) (see also Tab. 6.2). In order to illustrate the wide field of application of the chosen homogenization and discretization technique, ensembles containing 10, 30, 50, 70 and 80 volume percent of particles are investigated. A total of 50 different realizations was considered for each microstructure leading to 250 different finite element discretizations.

Mesh density study. In order to investigate convergence with respect to the mesh density, ensembles at 10, 20 and 30% volume fraction were examined. The microstructures are based on 10, 20 and 20 Voronoi generators, respectively. The results of the seven refinement levels h_i $(i = 0, \ldots, 6)$ of the microstructure are compared to the finest discretization h_7 in Tab. 6.3, Tab. 6.4 and Tab. 6.5. The finest mesh of each unit cell contains $\approx$1.1 mio. degrees of freedom ($\approx$360'000 nodes) and was chosen as a reference. As an error measure the largest and smallest proper values $\lambda_i^{\max}, \lambda_i^{\min}$ of $(\mathbb{C}_{h_7}^{\mathrm{a}})^{-1}\mathbb{C}_{h_i}^{\mathrm{a}}$ are analyzed to judge on convergence with respect to the mechanical properties, and $e_i^\theta = \|\bar{\boldsymbol{\alpha}}_{\theta,h_i}^{\mathrm{a}} - \bar{\boldsymbol{\alpha}}_{\theta,7}^{\mathrm{a}}\|/\|\bar{\boldsymbol{\alpha}}_{\theta,7}^{\mathrm{a}}\|$ is evaluated for the thermal expansion coefficient. Notably the difference between the finest and the coarser discretization is in the (sub-)percent regime. For the Monte Carlo type study a refinement level of approximately 100'000 nodes was chosen which is approximately equivalent to the refinement level h_2. Thereby, a relative precision of 1% is taken to be granted. Three refinement levels for the microstructure containing 10% of particles is shown in Fig. 6.7.

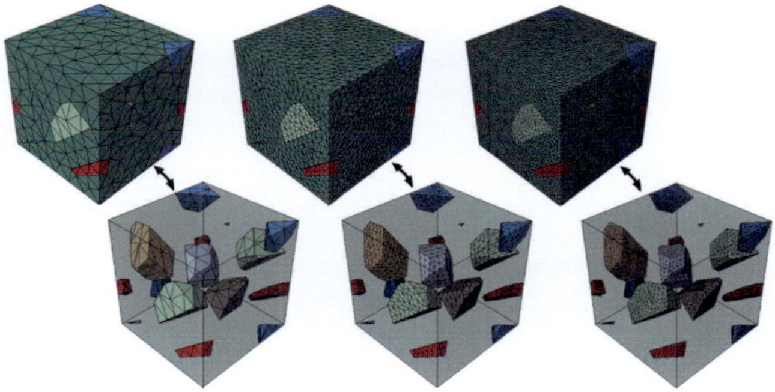

Figure 6.7: Different discretization levels 6502 (left), 40531 (middle) and 363808 (right) nodes for 10 particles (10 % vol. frac.) (see Fritzen and Böhlke, 2011c, for details)

refinement	h_0	h_1	h_2	h_3	h_4	h_5	h_6
# nodes	6'502	11'384	21'024	40'531	79'258	158'529	236'590
$\lambda_i^{\max} - 1$ [%]	0.8768	0.5844	0.3472	0.2066	0.1123	0.0512	0.0230
$\|\lambda_i^{\min} - 1\|$ [%]	0.1687	0.1103	0.0669	0.0396	0.0211	0.0093	0.0042
e_i^θ [%]	0.2172	0.1430	0.0870	0.0522	0.0279	0.0126	0.0062

Table 6.3: Results of the convergence study for particulate MMCs (10% volume fraction) (cf. to Fritzen and Böhlke, 2011c)

refinement	h_0	h_1	h_2	h_3	h_4	h_5	h_6
# nodes	10'626	14'363	22'909	41'385	79'191	155'374	232'162
$\lambda_i^{\max} - 1$ [%]	2.0628	1.4782	0.7912	0.4907	0.2630	0.1147	0.0529
$\|\lambda_i^{\min} - 1\|$ [%]	0.4217	0.2908	0.1654	0.0991	0.0519	0.0234	0.0108
e_i^θ [%]	0.5378	0.3742	0.2108	0.1265	0.0664	0.0299	0.0142

Table 6.4: Results of the convergence study for particulate MMCs (20% volume fraction) (cf. to Fritzen and Böhlke, 2011c)

refinement	h_0	h_1	h_2	h_3	h_4	h_5	h_6
# nodes	10'150	13'649	23'257	40'179	77'729	152'091	228'745
$\lambda_i^{\max} - 1$ [%]	2.3906	1.9991	1.1268	0.5878	0.3335	0.1396	0.0612
$\lvert\lambda_i^{\min} - 1\rvert$ [%]	0.4493	0.3684	0.1976	0.1143	0.0606	0.0256	0.0116
e_i^θ [%]	0.5907	0.4835	0.2633	0.1501	0.0796	0.0338	0.0153

Table 6.5: Results of the convergence study for particulate MMCs (30% volume fraction) (cf. to Fritzen and Böhlke, 2011c)

Influence of the particle volume fraction. The results of the computational homogenization for the 5 different particle volume fractions (10, 30, 50, 70, 80%) is presented in Tab. 6.6. Both, the Young's modulus and the Poisson ratio, are computed from the bulk and shear modulus based on (6.6) in order to provide engineering constants. Note that based on the truncated Taylor expansions

$$\bar{E}(\bar{K}+\delta K, \bar{G}+\delta G) \approx E(\bar{K},\bar{G}) + \partial_K \bar{E}(\bar{K},\bar{G})\delta K + \partial_G \bar{E}(\bar{K},\bar{G})\delta G, \tag{6.13}$$

$$\bar{\nu}(\bar{K}+\delta K, \bar{G}+\delta G) \approx \nu(\bar{K},\bar{G}) + \partial_K \bar{\nu}(\bar{K},\bar{G})\delta K + \partial_G \bar{\nu}(\bar{K},\bar{G})\delta G, \tag{6.14}$$

approximations of the respective standard deviations are given by

$$\sigma(E) \approx \sqrt{(\partial_K \bar{E}(\bar{K},\bar{G}))^2\sigma^2(K) + (\partial_G \bar{E}(\bar{K},\bar{G}))^2\sigma^2(G)}, \tag{6.15}$$

$$\sigma(\nu) \approx \sqrt{(\partial_K \bar{\nu}(\bar{K},\bar{G}))^2\sigma^2(K) + (\partial_G \bar{\nu}(\bar{K},\bar{G}))^2\sigma^2(G)}. \tag{6.16}$$

These approximations hold for small perturbation $\sigma(K), \sigma(G)$, as observed in the present study. In order to investigate the efficiency of analytical homogenization methods, the Voigt and Reuss bounds and the upper and lower Hashin-Shtrikman bounds are plotted together with the values obtained from the finite element analysis in Fig. 6.8. Additionally, a self-consistent approximation is also considered.

vol. frac.	10%	30%	50%	70%	80%
$\bar{K}$ [GPa]	75.50 ± 0.12	91.72 ± 0.22	112.8 ± 0.28	141.7 ± 0.27	160.7 ± 0.19
$\bar{G}$ [GPa]	31.21 ± 0.12	43.54 ± 0.19	61.18 ± 0.26	87.60 ± 0.34	106.1 ± 0.34
$\bar{E}$ [GPa]	82.29 ± 0.28	112.8 ± 0.44	155.4 ± 0.57	217.9 ± 0.71	260.9 ± 0.70
$\bar{\nu}$ [10^{-2}]	31.83 ± 0.06	29.51 ± 0.09	27.03 ± 0.09	24.37 ± 0.09	22.93 ± 0.07
$\bar{\alpha}$ [10^{-6}K^{-1}]	22.23 ± 0.04	17.35 ± 0.05	13.10 ± 0.04	9.328 ± 0.02	7.585 ± 0.01

Table 6.6: Effective isotropic thermo-elastic material parameters and 3σ confidence intervals for Al/SiC at different particle volume fractions (cf. to Fritzen and Böhlke, 2011c)

Influence of the average particle shape. The influence of the particle shape was investigated using modified Voronoi tessellations. Two types of modifications have been used: (i) the hard-core Voronoi tessellation (HC) and (ii) the centroidal Voronoi tessellation (CV). Details on the construction of the constraint tessellations are provided in section 3.3. In general, the particles resulting from constraint Voronoi generators are more regular, i.e. closer to being morphologically isotropic (see, e.g., Fig. 6.9).

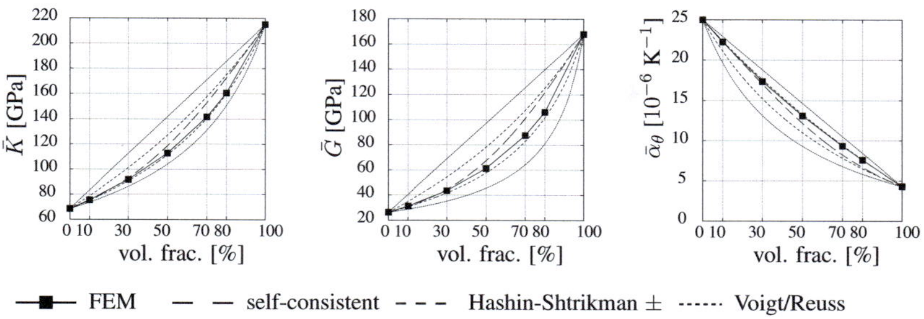

Figure 6.8: Comparison of the finite element results to analytical homogenization methods (cf. to Fritzen and Böhlke, 2011c)

Figure 6.9: Comparison of the particle shape obtained from an unconstraint Voronoi tessellation (left) and a centroidal Voronoi tessellation (right) (50 Voronoi generators, 30% particle volume fraction)

The same calculations as for the standard Voronoi tessellation (SV) have been carried out for two different repulsion distances (HC0.15, HC0.20) and the centroidal type of tessellation. This results in an additional 750 linear thermo-elastic calculations (seven load cases each). The results of this study are shown in Fig. 6.10 for the bulk and the shear modulus. In addition to the deviation in the effective value, the standard deviation has also been evaluated in order to compare the scattering of different realizations for the given microstructures. The values are normalized with respect to the results of the standard Voronoi tessellation (Fig. 6.11). The same comparison is carried out for the thermal expansion coefficient. The results are shown in Fig. 6.12.

Discussion. Periodic spatial discretizations of the model microstructures introduced in section 3.3.2 for materials with particle reinforcement have been used in a statistical study in order to estimate the effective thermo-mechanical properties of metal matrix composites. The considered material pairing (Al/SiC) is representative for a large variety of MMCs, e.g., for Al/Al_2O_3 or Al/B_4C. Particle volume fractions varying between 10 and 80% have

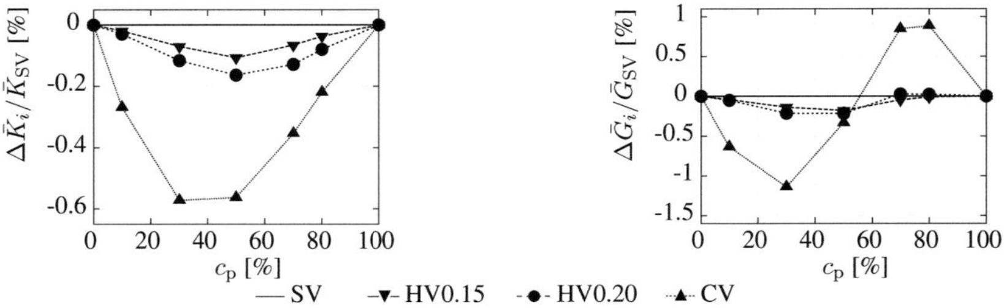

Figure 6.10: Comparison of the normalized deviation of the effective bulk $(\bar{K}_i - \bar{K}_{SV})/\bar{K}_{SV}$ and shear modulus $(\bar{G}_i - \bar{G}_{SV})/\bar{G}_{SV}$ for the different tessellations (HV0.15, HV0.20, CV) (left: bulk modulus K, right: shear modulus G) as a function of the particle volume fraction c_p (cf. to Fritzen and Böhlke, 2011c)

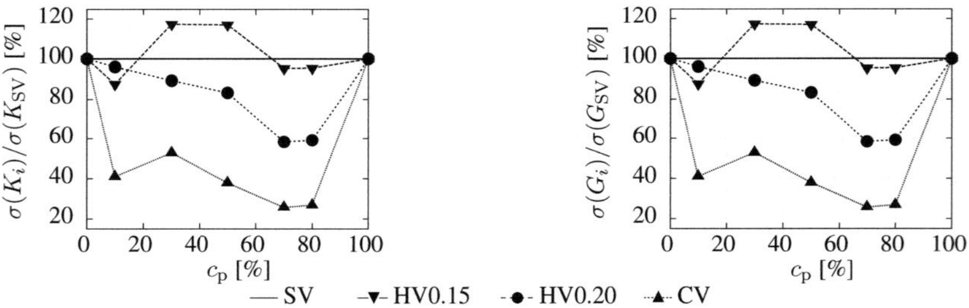

Figure 6.11: Comparison of the normalized standard deviations of bulk modulus $\sigma(K_i)/\sigma(K_{SV})$ (left) and shear modulus $\sigma(G_i)/\sigma(G_{SV})$ (right) for tessellation types HV0.15, HV0.20, CV over particle volume fraction c_p (cf. to Fritzen and Böhlke, 2011c)

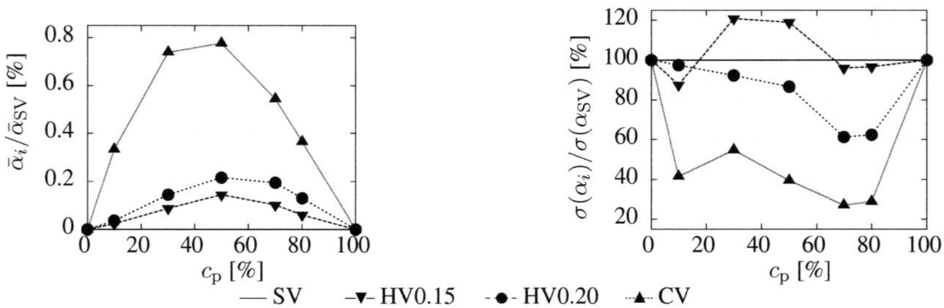

Figure 6.12: Comparison of the effective isotropic thermal expansion coefficient $\bar{\alpha}_i/\bar{\alpha}_{SV}$ (left) and its normalized standard deviation $\sigma(\alpha_i)/\sigma(\alpha_{SV})$ (right) for tessellation types HV0.15, HV0.20, CV over particle volume fraction c_p

been investigated. The method can hence be considered to be applicable to a wide range of composites with quasi-isotropic particle morphology.

For the considered class of microstructures a considerable agreement of the numerical results and the lower Hashin-Shtrikman estimate was found for all particle volume fractions. However, deviations of up to approximately 10% are found. The deviations are more pronounced for 30, 50 and 70% of particles. This range can be considered as important from a practical point of view, see e.g. Miserez et al. (2004) where microstructures containing 50-60% of particles are considered. The simple bounds (Voigt/Reuss estimate) are not satisfactory for the given contrast in the elastic moduli. Notably, the self-consistent approximation gives good results for 0-30 % particles. However, the effective shear modulus is overestimated for higher particle volume fractions. The results found for the elastic moduli can be transfered to the thermal expansion coefficient. Again the lower Hashin-Shtrikman bound gives a good prediction for all volume fractions with a relative error that can be considered negligible (sub-percent regime). This is due to the excellent agreement found for the effective bulk modulus (see Fig. 6.8, left) which is directly related to the effective thermal expansion coefficient via the Hashin-Rosen formula

$$\bar{\alpha}_\theta = (1 - c_\mathrm{p})\alpha_\mathrm{m} + c_\mathrm{p}\alpha_\mathrm{p} + \frac{\alpha_\mathrm{m} - \alpha_\mathrm{p}}{K_\mathrm{m}^{-1} - K_\mathrm{p}^{-1}} \left(\frac{1}{K^\mathrm{a}} - \frac{1 - c_\mathrm{p}}{K_\mathrm{m}} - \frac{c_\mathrm{p}}{K_\mathrm{p}} \right), \tag{6.17}$$

where K_p denotes the bulk modulus of the particles and K_m the bulk modulus of the matrix material and the thermal expansion coefficients $\alpha_\mathrm{p}, \alpha_\mathrm{m}$. While the effective properties $\bar{K}, \bar{\alpha}_\theta$ can thus be approximated by analytic methods, the results of the shear modulus should be verified in numerical calculations.

An interesting aspect for the proposed method is the possibility to examine arbitrary phase contrasts, e.g. ceramic particles embedded in a polymeric matrix. For these materials the analytic bounds span a prohibitively wide domain of possible material responses even for small volume fractions of only a few volume percent. In these cases numerical simulations should be considered in order to obtain more reliable data.

With respect to limitations of the used microstructures, the co-planar facets of neighboring particles and the rather uniform particle size distribution should be mentioned. Possible future modifications of the microstructure generation are easily possible. Suggestions in this direction include:

- The wall thickness parameter can be taken as a random variable with prescribed distribution. Thereby the particle size distribution and the nearest neighbor statistics can be influenced.
- An affine perturbation of the eroded cells in terms of a random translation and a random rotation of the particle can be introduced in order to manipulate the orientation of neighboring particle facets and the distribution of the nearest neighbor distance.

6.3 Inelastic homogenization of porous metals

In the following the inelastic material properties of porous metals are investigated in a Monte Carlo type computational study. An approximation of the microstructure of the porous material is given by the random model microstructures described in section 3.4, i.e. non-overlapping spherical voids in a periodic random arrangement are used. Pore volume fractions ranging from 0.1% to 30% are investigated.

6.3.1 Constitutive assumptions

In the following the mechanical behavior of porous metals is investigated. The constitutive behavior of the metallic material is assumed isotropic with the Young's modulus $E = 200$ GPa and the Poisson ratio $\nu = 0.3$. Inelastic incompressibility is asserted and a von Mises type plasticity model with constant yield stress $\sigma_{\mathrm{F}} = 100$ MPa is used. No hardening was considered in order to be able to relate the results to the ones predicted by existing analytical models in section 6.4. A linear kinematic description is employed with the infinitesimal strain tensor $\boldsymbol{\varepsilon}$ defined as the symmetric gradient of the displacement field $\boldsymbol{u}$. The Cauchy stress tensor is denoted $\boldsymbol{\sigma}$. The pores with boundary $\mathscr{P}$ are assumed to be pressure free. More specifically, the surface of the pores is a free boundary and the traction vector $\boldsymbol{t} = \boldsymbol{\sigma n}$ is zero on $\mathscr{P}$, where $\boldsymbol{n}$ denotes the unit normal pointing out of the pores.

A two-scale material is assumed with the smaller scale denoting the level at which individual pores are observed and the larger scale the structural level at which an effective (unknown) behavior is observed. The considered unit cell Ω_{tot} is taken as a cube in which the matrix $\Omega \subsetneq \Omega_{\mathrm{tot}}$ and the voids $\Omega_{\mathrm{tot}} \setminus \Omega$ are placed. The two scales are assumed to be clearly separated for usual homogenization theory to apply. The displacement field is assumed zero within the voids. In the given setting the microscopic and macroscopic stress and strain tensors are related by (see, e.g., Nemat-Nasser and Hori, 1999)

$$\bar{\boldsymbol{\varepsilon}} = \frac{1}{|\Omega_{\mathrm{tot}}|} \int_{\Omega} \boldsymbol{\varepsilon} \,\mathrm{d}V + \bar{\boldsymbol{\varepsilon}}_{\mathrm{c}}, \qquad \bar{\boldsymbol{\varepsilon}}_{\mathrm{c}} = \frac{1}{|\Omega_{\mathrm{tot}}|} \int_{\mathscr{P}} \mathrm{sym}(\boldsymbol{u} \otimes \boldsymbol{n}) \,\mathrm{d}A, \tag{6.18}$$

$$\bar{\boldsymbol{\sigma}} = \frac{1}{|\Omega_{\mathrm{tot}}|} = (1 - f)\langle \boldsymbol{\sigma} \rangle_{\Omega}, \tag{6.19}$$

where $\bar{\boldsymbol{\varepsilon}}_{\mathrm{c}}$ represents the cavity strain due to the deformation of the boundary of the pores and f denotes the pore volume fraction.

6.3.2 Boundary conditions

For porous metals the investigation of varying stress triaxiality ratios is of interest in order to find a pressure depend yield criterion for the homogenized porous material. Due to the constant yield stress on the small scale the application of stress boundary conditions, i.e., of anti-periodic traction boundary conditions, is not pursued due to the delicate assessment of admissible load increments. In order to still allow for varying stress triaxiality without constraining the deformation of the material too much a mixed periodic boundary condition was chosen. The variation of the triaxiality was performed using the real valued parameters α, β. For the simulation time t the boundary data for the unit cell calculations is

$$\left.\begin{aligned} \bar{\varepsilon}_{11}(t) &= t\dot{\varepsilon}_0(\alpha + \beta), \\ \bar{\varepsilon}_{22}(t) &= t\dot{\varepsilon}_0(-\alpha + \beta), \\ \bar{\varepsilon}_{33}(t) &= t\dot{\varepsilon}_0\beta, \\ \bar{\sigma}_{12}(t) &= \bar{\sigma}_{13}(t) = \bar{\sigma}_{23}(t) = 0. \end{aligned}\right\} \tag{6.20}$$

Here $\dot{\varepsilon}_0 > 0$ is a prescribed reference deformation rate. The given boundary conditions imply that the first invariant I and the second invariant J_2 of the macroscopic strain tensor are

$$I(\bar{\boldsymbol{\varepsilon}}) = \operatorname{tr}(\bar{\boldsymbol{\varepsilon}}) = 3\beta t\dot{\varepsilon}_0, \qquad J_2(\bar{\boldsymbol{\varepsilon}}) = \sqrt{3}|\alpha t\dot{\varepsilon}_0|. \tag{6.21}$$

Using the proposed loading, both, the macroscopic von Mises equivalent stress Σ_{eq} and the hydrostatic (or mean) stress Σ_{m}, can be controlled. The latter are defined via

$$\Sigma_{\text{eq}} = \sqrt{\frac{3}{2}}\|\bar{\boldsymbol{\sigma}}'\|_2, \qquad \Sigma_{\text{m}} = \frac{1}{3}\operatorname{tr}(\bar{\boldsymbol{\sigma}}), \tag{6.22}$$

where $\bar{\boldsymbol{\sigma}}'$ denotes the deviatoric part of $\bar{\boldsymbol{\sigma}}$. The nine different tupels (α_i, β_i) examined in the following are given in Tab. 6.7. Alternative stress-based loading conditions are often used to control the triaxiality ratio during loading (see, e.g., Besson, 2004). However, for the more than 500 simulations considered in this work a sort of mixed periodic boundary condition was chosen due to its computational robustness.

i	1	2	3	4	5	6	7	8	9
α_i	1.00	1.00	1.00	1.00	1.00	1.00	1.00	0.50	0.00
β_i	0.00	0.05	0.10	0.15	0.25	0.50	1.00	1.00	1.00

Table 6.7: Load parameters α, β used in the simulations

In the finite element simulations the periodicity of the displacement field was enforced using a linear relation between the displacement of the master and slave nodes on the surface of the unit cell. A parallel self-written finite element implementation was used to solve the physically nonlinear problem. Good numerical performance was achieved using the iterative solver HIPS (http://hips.gforge.inria.fr) for the large linear systems obtained from the discretization of the linearized constitutive equations. An implicit time integration procedure based on a backward Euler scheme was used. Due to the absence of hardening, convergence was hard to attain in all computations. A reason for this is found in the lack of strong ellipticity of the incremental Gibb's potential associated with the material law due to the lack of isotropic hardening. A line search procedure incorporating the Armijo rule was found to improve the numerical performance significantly, or to allow for convergence at all. This holds particularly true for the very small pore fractions which result in largely fluctuating stress, strain and displacement fields. The path-dependency of the solution is accounted for by a sufficiently fine discretization of the problem with respect to the simulation time.

6.3.3 Statistical properties of the microstructures

Three different statistical properties of the microstructures have been analyzed in order to examine the representativity of the obtained results: (i) the existence of pore clusters, (ii) the k-nearest neighbor distance and (iii) the two-point correlation of the pores. For the detection of pore clusters the algorithm used, e.g., by Bilger et al. (2005) was employed. In order to find pore clusters, the radius of the non-overlapping spherical pores is artificially augmented by a positive offset δR. Based on a collision detection algorithm chains of interconnected pores are then identified. The procedure is illustrated in Fig. 6.13.

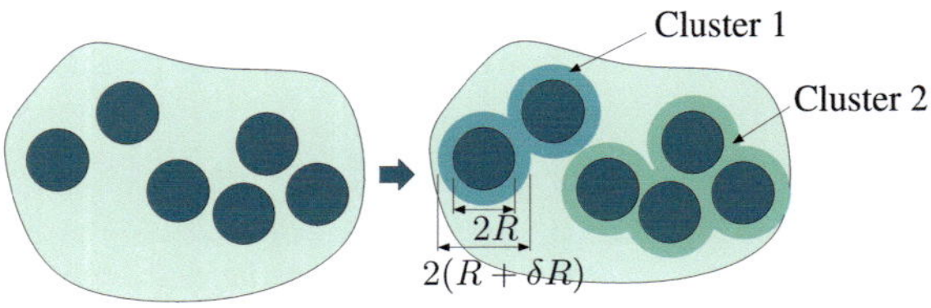

Figure 6.13: Schematic illustration of the pore cluster identification algorithm (two-dimensional case)

It was found that none of the microstructures showed pronounced clustering. However, almost all microstructures show micro-clusters consisting of 2 or 3 pores very close to each other (Fig. 6.14). The proposed algorithm is only of partial use for high volume fractions due to the small variation in the pore distance.

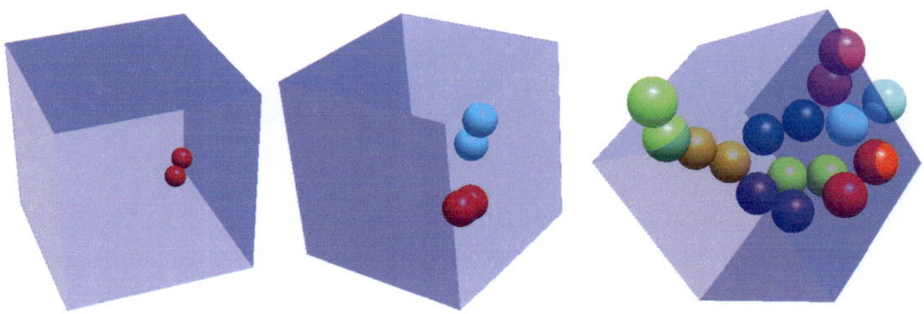

Figure 6.14: Pore clusters for $f = 1\%, N = 20$ ($\delta R = 0.3R$, left), $f = 5\%, N = 20$ ($\delta R = 0.2R$, middle), $f = 30\%, N = 50$ ($\delta R = 0.1R$, right); all shown pore clusters consist of 2 pores, except one at $f = 30\%$ (3 pores)

In order to analyze the average pore neighbor relationship of the microstructure, for the pore indexed i the k-nearest neighbor distance d_k^i is introduced as the minimum distance at which k neighbors are found ($1 \leq k < N$). The average value $\bar{d}_k$ over all pores and the standard deviation $\sigma(d_k)$ can be compared for different realizations of the microstructures to judge on the representativity. Other values of interest are the extremal magnitudes (min./max.) of d_k over all pores.

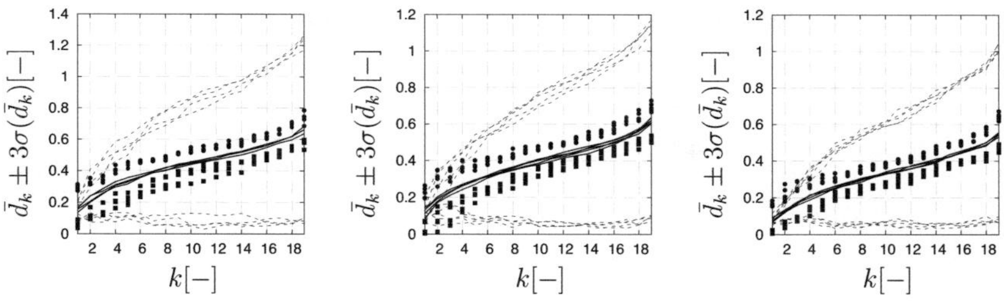

Figure 6.15: Average k-nearest distance of the pores for 0.1% (left), 1% (middle) and 5% pore volume fraction (right); 5 realizations with min./max. values (points), $k = 1, \ldots, 19$; dashed lines: 3σ confidence interval (values normalized with respect to the edge length of the unit cell)

In Fig. 6.15 the average value $\bar{d}_k$ is plotted for five realizations at different volume fractions (lines), where d_k is given as a non-dimensional value with respect to the length of an edge of the unit cell. Additionally, the minima and maxima are shown (squares and dots) and the $\pm 3\sigma(d_k)$ confidence interval is indicated in terms of dashed lines. For all volume fractions it is found that both the mean value and the standard deviation of d_k vary only slightly (see also the statistical data provided in section 3.4). However, some fluctuation in the minimum and maximum is found. Nevertheless, these fluctuations can be considered negligible. The results obtained for the other volume fractions are omitted for brevity. They show the same qualitative results. Due to the presented results the microstructures can be considered to show similar pore-pore neighbor relations.

The third statistical tool used to evaluate the properties of the microstructure is the two-point autocorrelation function of the pores. An algorithm based on geometrically exact intersection of spheres was developed and implemented into a C++ program. Multi-threading was used to reduce the computational time on shared memory multi-processor systems and near optimal speed-up was achieved. The autocorrelation function is plotted in Fig. 6.16 for two microstructures. It is found that the two-point function is almost directionally independent, i.e. nearly isotropic. As is known for models of non-overlapping circles and spheres, a region around the objects exists in which the two-point function is smaller than the asymptotic value f^2 at infinity. This property was also found for the examined microstructures.

6.3.4 Mesh density convergence study

In order to assure a sufficient quality of the presented results with respect to the spatial resolution, a convergence study has been performed on a unit cell for the parameters $f = 10\%, N = 10$. The load case $\alpha = 1, \beta = 0$ was chosen as an example. The finest solution was taken as reference and the response of the other discretizations is related to the last calculated value of this discretization in Fig. 3.40. The error with respect to the spatial

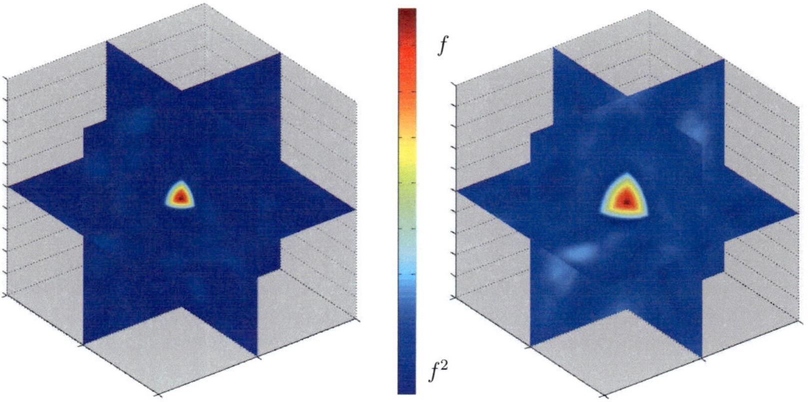

Figure 6.16: Two-point autocorrelation of the pores for $f = 1\%$ (left) and $f = 10\%$ (right)

discretization is found to decrease monotonically and the difference between two successive refinement levels tends to zero. With respect to the hydrostatic stress it should be noted that the reference value was only 3.3736 MPa and the absolute deviations for all meshes are smaller than 0.5 MPa which equals approximately 1% of $\Sigma_{\mathrm{eq}}^{\infty,h_6}$. Based on this study a medium mesh density was chosen for the statistical studies. Moreover, on the basis of the convergence study the error for the used discretizations is considered small with respect to other influence factors, such as the variation of the microstructural geometry between different realizations.

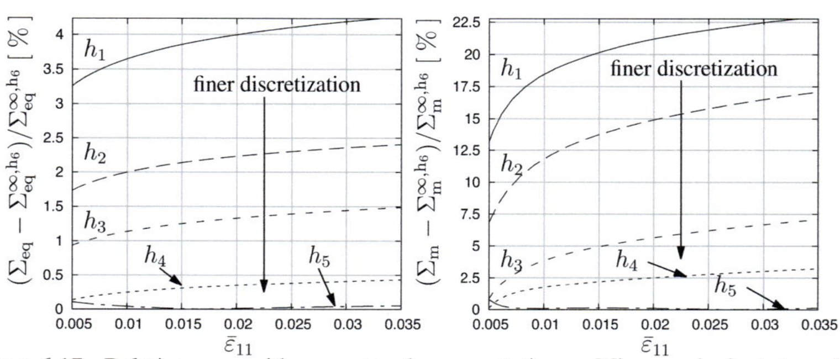

Figure 6.17: Relative error with respect to the asymptotic von Mises equivalent stress (left) and asymptotic effective hydrostatic stress (right) for the discretizations $h_1, \ldots, h_5$ (asymptotic value of the finest discretization taken as reference)

6.3.5 Asymptotic results

The asymptotic behavior of the porous microstructures is subject of the present investigation. In order to approximate the asymptotic response of the porous medium, the macroscopic stress at the end of the simulation time is used. It was verified that all of the taken values are stationary in the sense that the change in the macroscopic stress tends to zero when the overall strain loading is increased. In order to do so, the macroscopic von Mises equivalent and hydrostatic stress are plotted versus the non-dimensional loading time in Fig. 6.18. It is found that the stresses are stationary at the end of the simulations. This holds true for all examined microstructures.

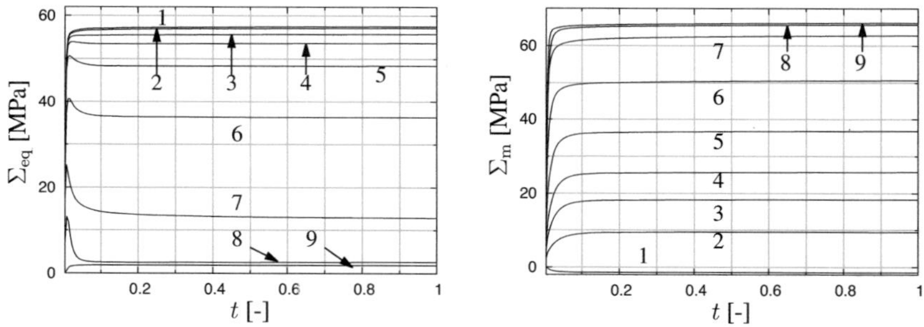

Figure 6.18: Effective von Mises equivalent stress (left) and hydrostatic stress (right) versus non-dimensional loading time for f =30% for all nine considered load cases (numbers indicate load case cf. Table 6.7)

The asymptotic behavior is plotted in the stress space (Σ_{eq} vs. Σ_m) in Fig. 6.19 for all examined microstructures and all nine considered loading directions. The results of the single pore model for f =0.1% and 30% are also shown. The computed data can be used to compare analytic models to simulation data and to identify parameters for phenomenological approaches.

6.3.6 Representativity of the results

By comparison of the single pore models and the random microstructures it can be concluded that for the dilute limit $f \to 0\%$, a single pore in a reference cell appears to give a sufficiently close approximation of the ductile behavior of the examined random isotropic microstructures. This is not the case for the higher volume fractions. The asymptotic stress response of examined microstructures at 30% pore volume fraction including 50 pores each shows a deviation of more than 10% from the single pore model.

In order to verify that the results obtained from the random microstructures are representative for the different volume fractions, the asymptotic stresses obtained from the statistical study at a porosity of 20% and using 50 pores are examined closely for the loading path

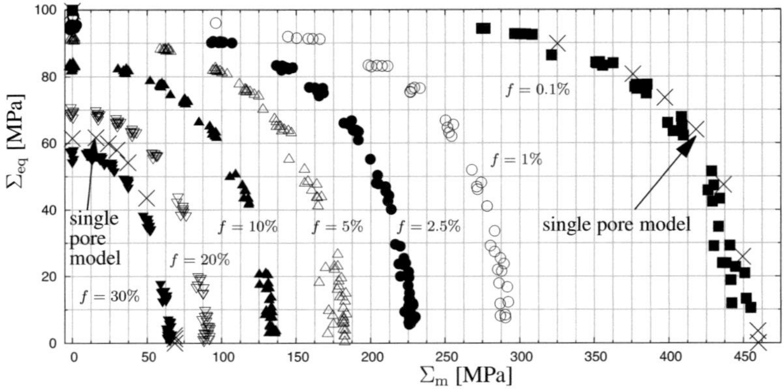

Figure 6.19: Calculated asymptotic stress response of the porous microstructures for pore volume fractions from 0.1 to 30%; the single-pore model is shown for comparison at f =0.1% and 30%

characterized by $\alpha = 1, \beta = 0.25$. The latter was chosen because it shows both, pronounced deviatoric and hydrostatic stress components. Three different points are investigated: (i) the convergence with respect to the number of pores, (ii) the convergence with respect to different types of boundary conditions and (iii) convergence with respect to the ensemble average of the Monte Carlo study.

Therefore, a new set of random microstructures was created and discretized for f =20% and varying number of pores ($N \in \{5, 10, 20, 50, 75, 100, 150, 200\}$). All these microstructures are subjected to the indicated loading path and the asymptotic stresses are analyzed. In addition to the periodic boundary conditions described before, kinematic uniform boundary conditions were applied to the same cells. A minor modification was necessary in the sense that the entire displacement on the boundary was prescribed which slightly differs from the mixed periodic boundary conditions described in section 6.3.2. It was verified that both, the effective shear stresses with kinematic uniform boundary conditions and the effective shear strains with periodic boundary conditions are close to zero, i.e. that the influence of this difference can be considered negligible. Indeed, for a representative microstructure, i.e. for an infinite number of pores, the response given by the different boundary conditions should be identical. In order to guarantee independence of the fluctuations with respect to the chosen spatial discretization, rather fine meshes with up to one million degrees of freedom are used.

The results are shown in Fig. 6.20. The difference between the two loading conditions for the same realization is rather pronounced (box and circles) at a low number of pores. This does particularly hold true for the hydrostatic part of the stress (Fig. 6.20, right). For higher pore numbers the differences due to the boundary condition reduce significantly.

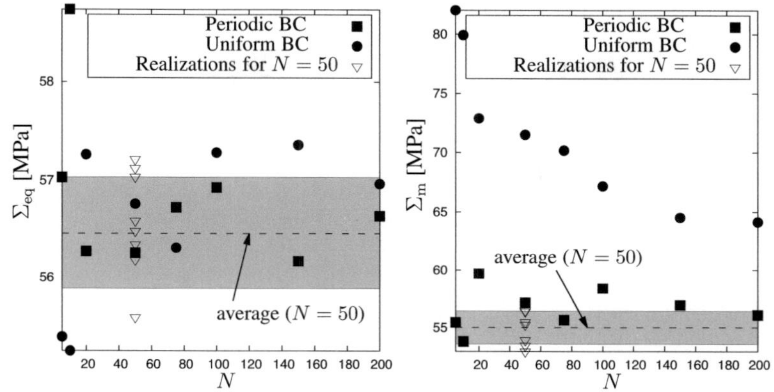

Figure 6.20: Comparison of the asymptotic von Mises equivalent stress (left) and hydrostatic stress (right) for different numbers of pores and boundary conditions; the values of the statistical study are also shown (gray area: $\pm\sigma(\Sigma_{eq})$ and $\pm\sigma(\Sigma_m)$)

Moreover, the average value obtained from the statistical study at N =50 is plotted in terms of a dashed line and the confidence interval of the statistical study given by plus/minus the standard deviation is indicated in gray. Notably, all computed points with periodic boundary conditions are found within or close to this range. Additionally, the asymptotic values for periodic boundary data for the largest unit cell (N =200) do almost coincide with the averages from the statistical investigation. More precisely, the realizations containing 150 or 200 pores show a deviation of less than 1% in the von Mises equivalent stress and 2% in the hydrostatic stress from this line. The mentioned points show that the computational results presented earlier are representative for the respective pore fractions.

Moreover, the average value obtained from the statistical study at N =50 is plotted in terms of a dashed line and the confidence interval of the statistical study given by plus/minus the standard deviation is indicated in gray. Notably, all computed points with periodic boundary conditions are found within or close to this range. Additionally, the asymptotic values for periodic boundary data for the largest unit cell (N =200) do almost coincide with the averages from the statistical investigation. More precisely, the realizations containing 150 or 200 pores show a deviation of less than 1% in the von Mises equivalent stress and 2% in the hydrostatic stress from this line. Interestingly, the effective von Mises equivalent stress for the uniform boundary conditions is rather close to the solution obtained for the periodic displacement fluctuation boundary conditions. However, the mean stress is grossly overestimated.

Surprisingly some of the effective stress responses of the UKBC are softer than for the PKBC, see for instance Σ_{eq} for $N = 5$ or $N = 10$, although they are known to provide an upper bound for the PKBC. This is due to the fact that the bounding character is not found in terms of the stresses, but in in terms of the work per unit volume needed to attain the final prescribed macroscopic deformation. The latter is computed for the time interval $[0, T]$ by virtue to the Hill-Mandel condition

$$\bar{W}_{\text{tot}} = \int_0^T \langle \boldsymbol{\sigma} \cdot \dot{\boldsymbol{\varepsilon}} \rangle \, \mathrm{d}t = \int_0^T \bar{\boldsymbol{\sigma}} \cdot \dot{\bar{\boldsymbol{\varepsilon}}} \, \mathrm{d}t. \tag{6.23}$$

The total work has been calculated for both types of boundary conditions and for all considered values of N in Fig. 6.21. While the values found for the periodic solutions is almost identical for all values of N, i.e., even for $N = 5$, the uniform boundary conditions show significantly higher values. Further, it is found that for homogeneous displacements prescribed on the boundary the energy is decreasing with rising N.

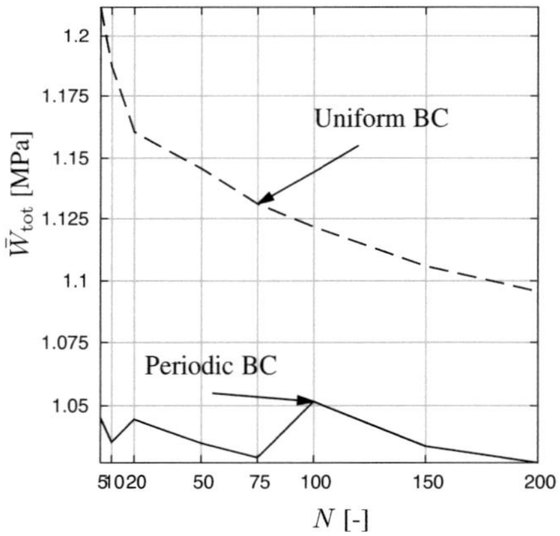

Figure 6.21: Comparison of the work density $\bar{W}_{\text{tot}}$ needed to attain the prescribed deformation for uniform kinematic boundary conditions and periodic fluctuation fields

In addition to the analysis of the effective stress response and the expended work, the local plastic strain fields are also analyzed. For $N = 5$ (Fig. 6.22) and $N = 200$ (Fig. 6.23) the computed equivalent plastic strain at the end of the simulation is compared. On the left hand side the uniform boundary conditions are shown, on the right hand side the field obtained using periodic displacement fluctuations is plotted. For a small number of pores the plastic strain fields are largely different for the two types of loading. For $N = 200$ the effect is limited to a boundary close region, i.e., the plastic strain found close to the surface is significantly different. However, the fields found inside of the cell show an excellent agreement. This leads to the conclusion that for an infinite cell size the two solutions are expected to coincide almost everywhere with respect to the Borel-Lebesgue measure. Moreover the discrepancies still observed for 200 pores show that the needed cell size to attain this result is presumingly very large, i.e. it involves thousands of spherical voids.

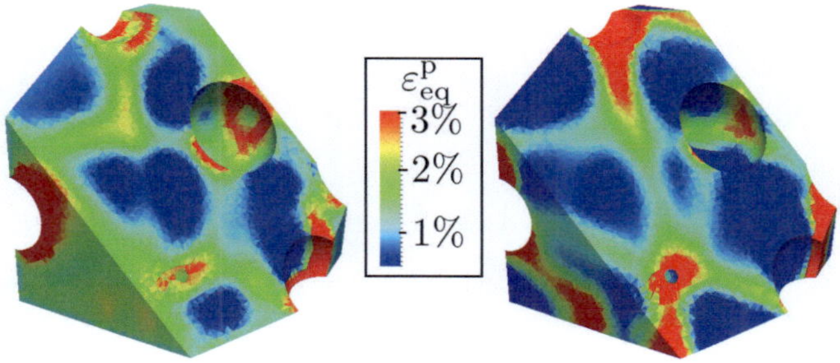

Figure 6.22: Accumulated plastic strain for UKBC (left) and PKBC (right) at the final prescribed loading for the microstructure containing 5 pores

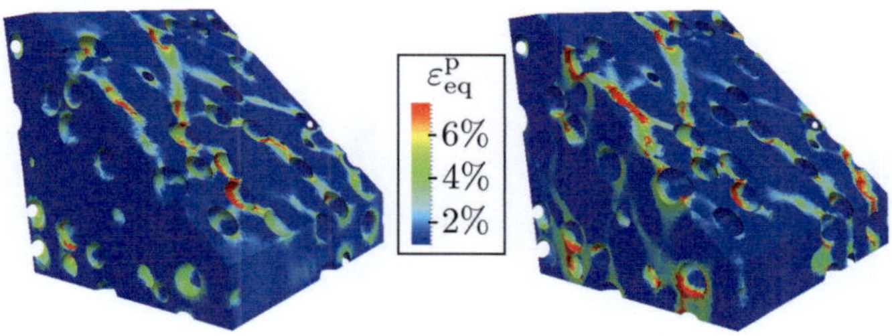

Figure 6.23: Accumulated plastic strain for UKBC (left) and PKBC (right) at the final prescribed loading for the microstructure containing 200 pores

The mentioned results show that the computational results presented earlier can be considered representative for the considered microstructural class.

6.3.7 Local plastic strain fields

The distribution of the accumulated plastic strain field has been analyzed for different loading conditions and microstructures. Perfectly plastic, i.e. non-hardening, metals tend to form strongly localizing bands. For instance, Bilger et al. (2005) reported a pronounced influence of the applied hydrostatic load on the dispersion and curvature of the localization zones of the plastic deformation for periodic unit cells containing 39 pores at a volume fraction of 0.6%. The results of the present study confirm these observations for all examined volume fractions, see for instance Fig. 6.24.

In addition it was found that for the higher volume fractions the plastic strain tends to be more diffuse than for the small volume fractions. The resulting patterns are complex due to the small distance between neighboring pores with respect to their diameter (Fig. 6.25). At a porosity of 30%, the diameter of a single pore in an aggregate containing 50 voids is 0.2254 of the edge length of the unit cell. From the analysis of the k-nearest neighbor distance it

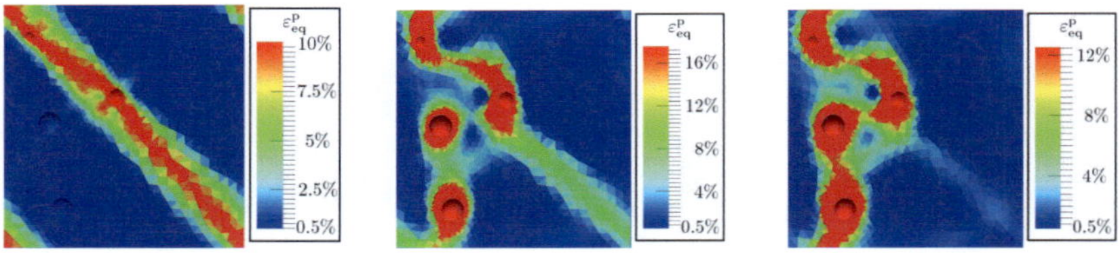

Figure 6.24: Distribution of the equivalent plastic strain at the end of the simulation for three load cases with $\alpha = 1$ and $\beta = 0$ (left), 0.25 (middle) and 0.50 (right) for a porosity of 1%

is found that within this distance 20 or more pores can be found for all of the considered realizations. Accordingly, a large amount of pore-pore interactions occur. For the smallest two porosities, at most one neighboring pore is found within a distance of $2R$. The amount of interaction is, thus, significantly reduced. This is already suggested by the two-point correlation function (Fig. 6.16).

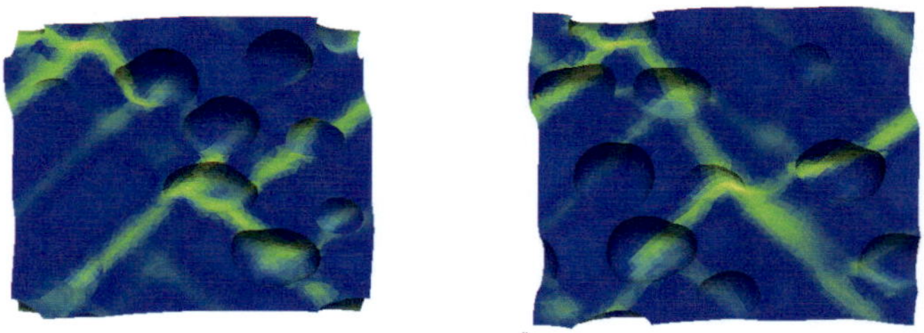

Figure 6.25: Distribution of the accumulated plastic strain in a high porosity material (30%); note the small deflections of the localization bands from 45° due to close neighboring pores (load case: $\alpha = 1, \beta = 0$); two slices of the entire volume are shown; the deformation is scaled by a factor of 2;

6.4 Interpretation and discussion

6.4.1 Comparison with analytical models

The computational results shown in Fig. 6.19 have been compared to some of the many existing analytical approaches. Besides the classical Gurson model (Gurson, 1977), the Gurson Hashin-Shtrikman upper bound (Leblond et al., 1994; Garajeu and Suquet, 1997), the Gărăjeu-Suquet upper bound (Garajeu and Suquet, 1997), the approach of Monchiet et al. (2007) and the approximation proposed by Tvergaard (1981) were chosen. The latter is a phenomenological extension of the original Gurson model by introduction of three coefficients $q_1, q_2, q_3 > 0$. The model is commonly referred to as Gurson-Tvergaard-Needleman

(GTN) model. The relation between the macroscopic von Mises equivalent stress and the hydrostatic stress is given by

$$\frac{\Sigma_{\rm eq}}{\sigma_{\rm F}} = \sqrt{1 + q_3 f^2 - 2 q_1 f \cosh\left(q_2 \frac{3\Sigma_{\rm m}}{2\sigma_{\rm F}}\right)}. \tag{6.24}$$

Notably, the Gurson model is recovered by setting $q_1 = q_2 = q_3 = 1$. Different values for the coefficients are postulated in the literature (Besson, 2004) with $q_1 = 1.25 - 2$, $q_2 \approx 1$ and $q_3 = q_1^2$. When trying to adjust the coefficients q_1, q_2, q_3 to the computed data it was immediately found that no set of parameters was able to predict all of the calculated asymptotic stress responses, see for instance Fig. 6.26 where the high pore volume fractions differ significantly from the model predictions. Additionally, $q_2 = 1$ was found not to reproduce the actual curves with a sufficient accuracy. In particular, the transition from the edgy shape for $f = 0.1\%$ to the smooth curve for high porosities (≥10%) is not possible.

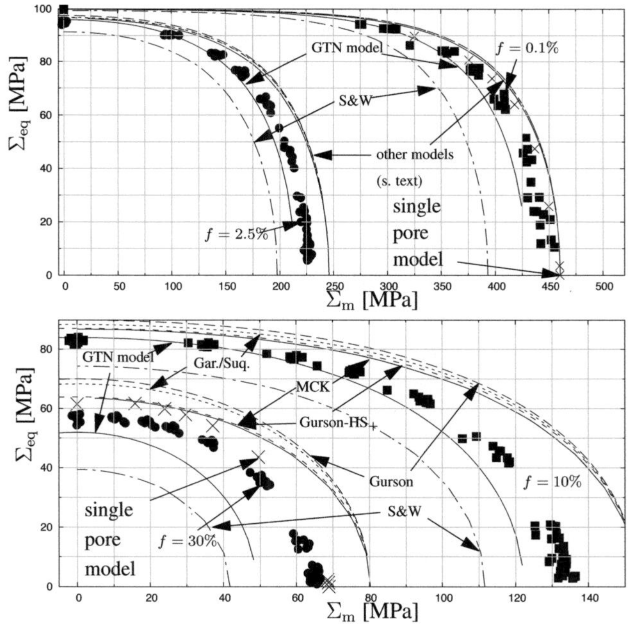

Figure 6.26: Comparison of the GTN ($q_1 = 1.5, q_2 = 1, q_3 = q_1^2$), the Gurson-HS estimate (Leblond et al., 1994; Garajeu and Suquet, 1997), the Garajeu-Suquet upper bound (Garajeu and Suquet, 1997), the model by Monchiet et al. (2007) (MCK) and the classical Gurson model with the computational results for $f = 0.1\%$, 2.5% (top) and 10%, 30% (bottom); the lower bound of Sun and Wang (1989) ($b_4 = 0$; S&W) is also shown

For the smaller porosity it was found that all of the other models show almost identical behavior (see Fig. 6.26, top), except for the Gurson model which overestimates the stresses and the GTN model which is a little closer to the computed values. The consideration of high porosities (10-30%, Fig. 6.26, bottom) reveals some differences between the different approaches. In summary, the stresses are overpredicted by all models, except the GTN model. The latter does underestimate the stress response for the coefficients determined at the low volume fractions. Note that all computed points satisfy the lower bound of Sun and Wang (1989), where $b_4 = 0$ was chosen since for $b_4 > 0$ the predicted stresses are even lower. For low pore volume fractions the bound appears to provide some useful estimates. For higher porosities the prediction of the stresses is much too soft in comparison with the computational results and the other analytical models.

6.4.2 Identification of a modified GTN model

In order to overcome the deficiency of the tested models, an adjusted GTN model is proposed based on the presented computational results. Evidently the coefficients q_1, q_2, q_3 are not constant for all volume fractions. It is found that the simple relations

$$q_1(f) = \theta_0 - \theta_1 f, \qquad q_2(f) = \theta_2, \qquad q_3(f) = (q_1(f))^2 \tag{6.25}$$

are sufficient to reproduce the results for the higher volume fractions while the results for the low porosities are still approximated to an excellent extent (Fig. 6.27). The three scalar coefficients identified based on the computational study are $\theta_0 = 1.69, \theta_1 = 1, \theta_2 = 0.92$. The value $\theta_1 = 1$ is remarkable.

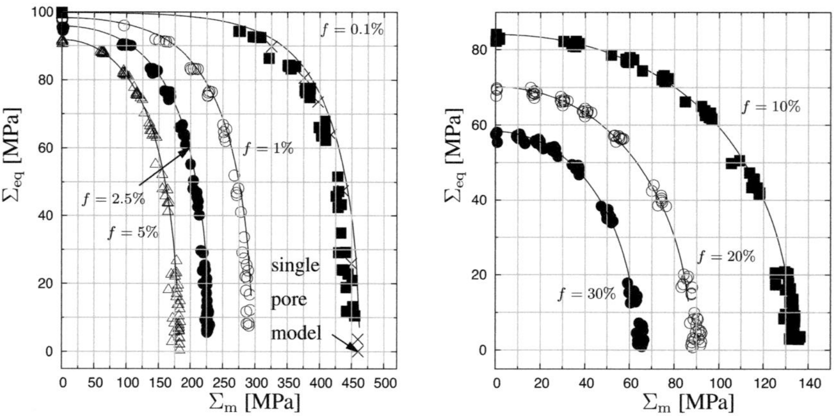

Figure 6.27: Adjusted GTN model and computational results

In order to find out the significance of the obtained asymptotic stress response with respect to the loading history, the time history of simulations at 1 and 30% pore volume fraction is shown in Fig. 6.28 together with the adjusted GTN model. It is found that the simulations do not describe a straight line in the particular choice of stress space but a rather complex non-proportional path. Interestingly, all of the calculated trajectories lie close to the adjusted GTN model, i.e. their position varies almost only tangential to the curve. These results suggest that the proposed model can describe the inelastic response of the structure at all stages of plastic loading and not only in the limit case of an asymptotic response.

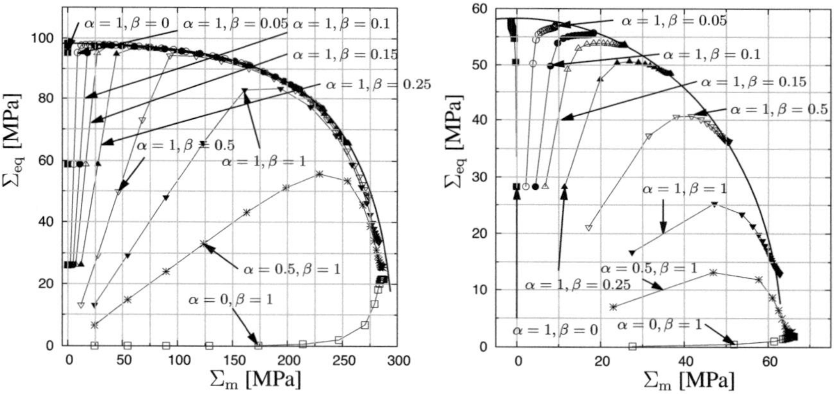

Figure 6.28: Time history of the nine investigated load cases for $f = 1\%$ (left) and $f = 30\%$ (right)

6.5 Non-uniform transformation field analysis of metal matrix composites

The NTFA presented in detail in chapter 5 is applied to the model microstructures presented in section 3.3.2. The aim is the approximation of the macroscopic constitutive response of metal matrix composites with particulate reinforcement. While the linear properties of these materials can be considered to be sufficiently well understood (see for instance section 6.2, Fritzen and Böhlke (2011c)), the effective material behavior in the presence of physical nonlinearities is subject of many investigations, see for instance the review article of Kanoute et al. (2009) or the introductory section of chapter 5. In the following the computational efficiency of the NTFA for microheterogeneous materials undergoing complicated loading is investigated. In particular non-uniform and multi-axial loading paths are considered and the homogenized model is used in large scale structural computations.

6.5.1 Considered microstructures

Microstructure A
The first examined microstructure contains 18.2 volume percent of elastic particles (10 particles) embedded into an elasto-plastic matrix material. The periodic spatial discretization is based on the approach presented in section 3.3.2 (see also Fritzen and Böhlke, 2011c). A representation of the mesh is shown in Fig. 6.29.

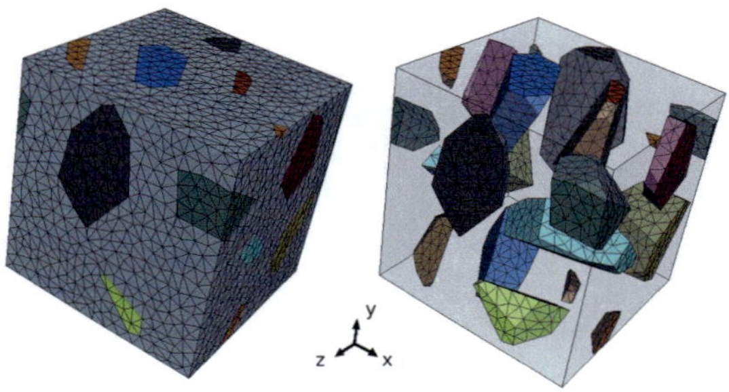

Figure 6.29: Spatial discretization of microstructure A (73'130 nodes, 52'473 elements) (cf. Fritzen and Böhlke, 2010b)

For the particles the linear elastic properties of Boron as reported by Michel and Suquet (2003) with the Young's modulus $E = 400$ GPa and the Poisson ratio $\nu = 0.2$ are used. The matrix material is assumed to have the elastic parameters $E = 75$ GPa and $\nu = 0.3$ associated with aluminum. The inelastic behavior of the metallic matrix material is considered rate-independent. A von Mises type plasticity law with nonlinear hardening described by

$$r(q) = \sigma_{\mathrm{F0}} + hq^{p} \tag{6.26}$$

is asserted. The parameters $\sigma_{F0} = 75$ MPa, $h = 416.5$ MPa and $p = 0.3895$ representing aluminum are decided on (cf. Michel and Suquet, 2003). The high degree of nonlinearity of the yield stress is depicted in Fig. 6.30.

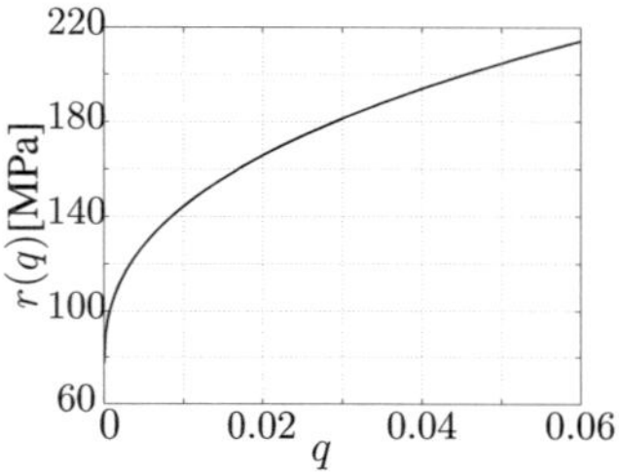

Figure 6.30: Yield stress $r(q)$ versus hardening parameter q for the polynomial hardening law (6.26) (cf. Fritzen and Böhlke, 2010b)

The mode identification is performed using an orthogonalization procedure proposed by Fritzen and Böhlke (2010b) (see also section 5.5) that results in a total of five inelastic modes for the considered microstructure. The mode identification was performed based on a total of five inelastic computations. The considered loading directions are

$$\begin{aligned}
&\dot{\bar{\varepsilon}}_1 = \dot{\varepsilon}_0(2\boldsymbol{e}_1 \otimes \boldsymbol{e}_1 - \boldsymbol{e}_2 \otimes \boldsymbol{e}_2 - \boldsymbol{e}_3 \otimes \boldsymbol{e}_3), \qquad &&\dot{\bar{\varepsilon}}_2 = \dot{\varepsilon}_0(\boldsymbol{e}_2 \otimes \boldsymbol{e}_2 - \boldsymbol{e}_3 \otimes \boldsymbol{e}_3),\\
&\dot{\bar{\varepsilon}}_3 = \dot{\varepsilon}_0(\boldsymbol{e}_1 \otimes \boldsymbol{e}_2 + \boldsymbol{e}_2 \otimes \boldsymbol{e}_1), &&\dot{\bar{\varepsilon}}_4 = \dot{\varepsilon}_0(\boldsymbol{e}_1 \otimes \boldsymbol{e}_2 + \boldsymbol{e}_2 \otimes \boldsymbol{e}_1),\\
&\dot{\bar{\varepsilon}}_5 = \dot{\varepsilon}_0(\boldsymbol{e}_1 \otimes \boldsymbol{e}_2 + \boldsymbol{e}_2 \otimes \boldsymbol{e}_1).
\end{aligned} \tag{6.27}$$

The load amplitude at the end of the simulation time was controlled by the rate $\dot{\varepsilon}_0$ such that a macroscopic strain of approximately 5% could be achieved in all computations.

Microstructures B.I, B.O2/3/4, B.E2/3/4

Following the proposal for a morphological anisotropy of the material in section 3.3.1.2 for polycrystalline aggregates, an anisotropic modification of the particles is used in Fritzen and Böhlke (2011b). Two classes of transversely isotropic affine transformations with three different amounts of anisotropy are applied to a structure containing 40 particles (20 % volume fraction) embedded into an elasto-plastic matrix material as for microstructure A. Example discretizations of the periodic microstructures are shown in Fig. 6.31.
For the inclusion phase the properties of SiC proposed by Chawla et al. (2006) are taken ($E = 400$ GPa, $\nu = 0.19$). The affine transformations describing the transversal anisotropy are ($\lambda \geq 1$)

$$\bar{\boldsymbol{F}}_{\rm o}(\lambda) = \begin{pmatrix} \lambda^{-1} & 0 & 0 \\ 0 & \lambda^{1/2} & 0 \\ 0 & 0 & \lambda^{1/2} \end{pmatrix}, \qquad \bar{\boldsymbol{F}}_{\rm e}(\lambda) = \begin{pmatrix} \lambda & 0 & 0 \\ 0 & \lambda^{-1/2} & 0 \\ 0 & 0 & \lambda^{-1/2} \end{pmatrix}. \tag{6.28}$$

The transformation $\bar{\boldsymbol{F}}_{\rm o}(\lambda)$ represents oblate microstructures and $\bar{\boldsymbol{F}}_{\rm e}(\lambda)$ results in elongated configurations. The three different values two, three and four are considered for λ for both,

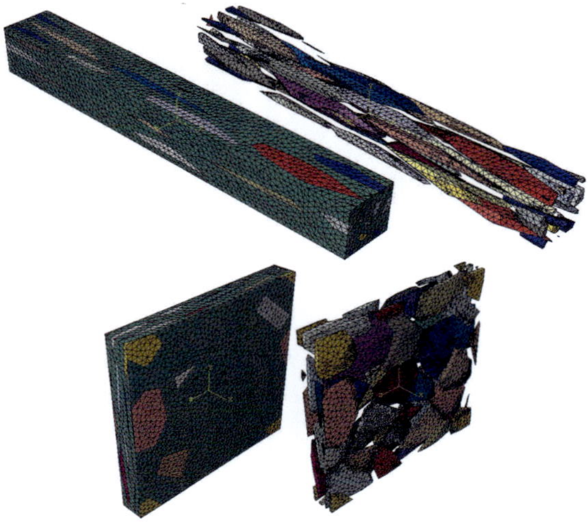

Figure 6.31: Finite element mesh of the microstructures B.E4 (top) and B.O4 (bottom) (cf. Fritzen and Böhlke, 2011b)

the oblate and the elongated morphology, in addition to the isotropic structure (i.e., $\lambda = 1$). In order to investigate the actual anisotropy, the elongation of each particle in x-, y- and z-direction is computed. For all microstructures the aspect ratio

$$\mathcal{A}_\mathrm{o} = \frac{l_y + l_z}{2 l_x}, \qquad \mathcal{A}_\mathrm{e} = \frac{2 l_x}{l_y + l_z}, \tag{6.29}$$

of the individual grains is then averaged over all inclusions. The results are shown in Tab. 6.8. Notably the grains are highly anisotropic with mean grain aspect ratios of up to 11.58. The oblate microstructures are labeled B.O2, B.O3 and B.O4. Analogously the names B.E2, B.E3, B.E4 and B.I are defined for the elongated and the isotropic microstructures, respectively. The numerical testing of the microstructures was performed based on the same kinematic loadings applied in the identification process of microstructure A.

	oblate				elongated		
λ	2	3	4		2	3	4
$\mathcal{A}_\mathrm{o}$	3.852	7.639	11.581	$\mathcal{A}_\mathrm{e}$	3.343	6.338	9.849

Table 6.8: Average particle aspect ratios for the oblate and the elongated microstructures (cf. Fritzen and Böhlke, 2011b)

6.5.2 Comparison to full-field simulations on integration point level

Results for microstructure A

The mode identification procedure used for microstructure A is based on the algorithm developed by Fritzen and Böhlke (2010b) which is described in section 5.5.3. The coupled model with the modified normalization condition (M2b) was used. Here $c = 0.818$ denotes the matrix volume fraction. In order to assess the capability of the NTFA model to accurately predict the stress response of the microheterogeneous material, the isothermal NTFA model and the corresponding fully resolved finite element simulation are compared. A non-proportional strain path of the form

$$\dot{\bar{\varepsilon}} = \begin{cases} \dfrac{1}{100\mathrm{s}}(\boldsymbol{e}_1 \otimes \boldsymbol{e}_1 - \boldsymbol{e}_2 \otimes \boldsymbol{e}_2) & 0 \leq t \leq 1\mathrm{s}, \\ \dfrac{1}{200\mathrm{s}}(1\boldsymbol{e}_2 \otimes \boldsymbol{e}_2 - 1\boldsymbol{e}_3 \otimes \boldsymbol{e}_3 + 4(\boldsymbol{e}_1 \otimes \boldsymbol{e}_3 + \boldsymbol{e}_3 \otimes \boldsymbol{e}_1)) & 1 < t \leq 2\mathrm{s}, \end{cases} \tag{6.30}$$

is prescribed to both problems, where the reduced model has five independent mode coefficients and the full-field simulation incorporates approximately 200'000 degrees of freedom and the additional internal variables. The effective stress tensor of both computations are compared, where the numerical full-field solution is considered as the reference. A partial load reversal was enforced by purpose to investigate the behavior of the model under pronounced load path changes.

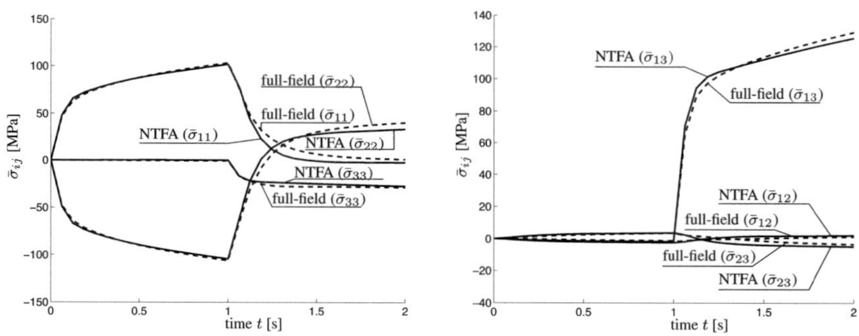

Figure 6.32: Time history of $\bar{\sigma}_{11}, \bar{\sigma}_{22}, \bar{\sigma}_{33}$ (left) and $\bar{\sigma}_{12}, \bar{\sigma}_{13}, \bar{\sigma}_{23}$ (right) (cf. Fritzen and Böhlke, 2010b)

All six components of the macroscopic stress tensor are compared in Fig. 6.32. An excellent agreement was found for all components, although the loading did not match any of the loading directions prescribed during the identification procedure. The computational effort for the solution of the homogenized material law is found in the range of sub-milliseconds per load increment, whereas the full-field simulation requires substantial computational resources in terms of both, CPU time and memory. It can be considered to take several hours, such that the order of computational saving is approximately 10^6. Interestingly, the computational effort for the homogenized material depends only on the number of inelastic modes, i.e., it is completely decoupled from the mesh density of the reference computations that are used to identify the inelastic modes. Two of the modes are shown in Fig. 6.33 to illustrate the heterogeneity of the local fields.

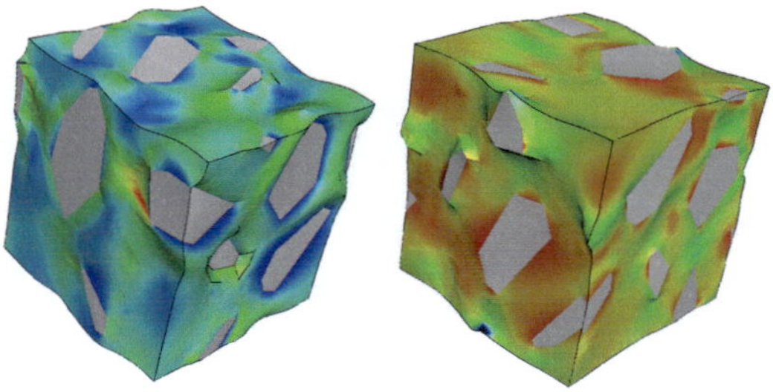

Figure 6.33: Plastic strain components 11 of mode 1 (left) and 33 for mode 2 (right) (cf. Fritzen and Böhlke, 2010b)

In addition to the comparison of the global stress response, a relocalization of the fields has been performed for the non-proportional loading path. The full-field computation and the fields reconstructed from the homogenized material law are compared in Fig. 6.34, 6.35.

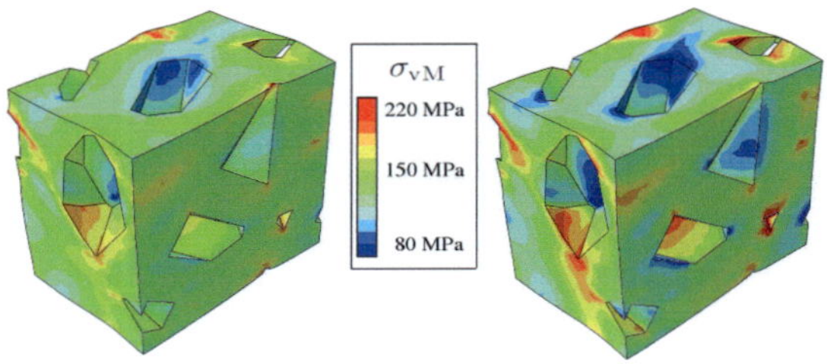

Figure 6.34: Comparison of the von Mises equivalent stress for the full-field simulation (left) and the relocalization (right) at the point of the load reversal (cf. Fritzen and Böhlke, 2010b)

A good agreement is found for the von Mises equivalent stress. More precisely, the positions of minimum and maximum stresses are accurately predicted although some quantitative deviations can be found.

Results for microstructures B.I, B.O2/3/4, B.E2/3/4

For the evaluation of the quality of the approximation of the elongated and oblate microstructures of type B, two different random loadings have been considered. In this study the modes

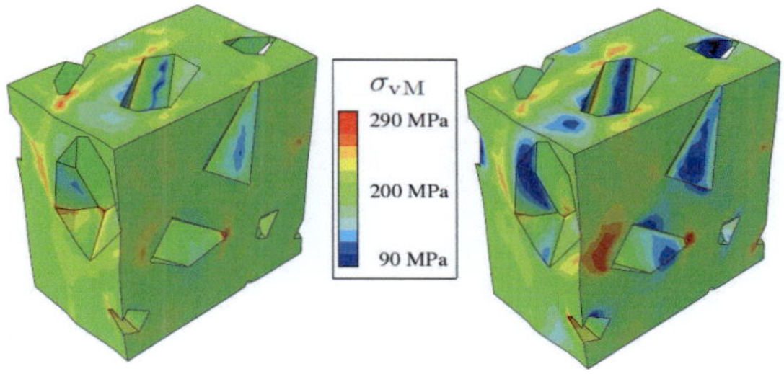

Figure 6.35: Comparison of the von Mises equivalent stress for the full-field simulation (left) and the relocalization (right) at the end of the simulation (cf. Fritzen and Böhlke, 2010b)

are identified based on the Karhunen-Loève decomposition presented in section 5.5. Between 11 and 13 modes have been found for the relative threshold $\delta = 10^{-5}$ relating the largest eigenvalue and the last accepted eigenvalue of the matrix $\hat{K}_{\rm p}$. After the system matrices have been determined the intial macroscopic yield surface of the material has been examined to demonstrate the anisotropy of the effective inelastic behavior for the elongated and the oblate microstructure in Fig. 6.36. The influence of the anisotropic particles onto the inelastic material response is immediately visible from the presented graph.

Figure 6.36: Initial macroscopic yield surface of the elongated (left) and oblate (right) microstructure for $\lambda = 4$

For the inelastic testing the two considered loadings are

$$\dot{\boldsymbol{\varepsilon}}_1 = 1\,{\rm s}^{-1} \begin{pmatrix} 0.0169 & -0.0405 & -0.0053 \\ & -0.0159 & -0.0103 \\ \text{sym.} & & 0.0392 \end{pmatrix} \boldsymbol{e}_i \otimes \boldsymbol{e}_j, \tag{6.31}$$

$$\dot{\boldsymbol{\varepsilon}}_2 = 1\,{\rm s}^{-1} \begin{pmatrix} 0.0361 & -0.0014 & -0.0074 \\ & 0.0401 & 0.0199 \\ \text{sym.} & & -0.0426 \end{pmatrix} \boldsymbol{e}_i \otimes \boldsymbol{e}_j. \tag{6.32}$$

Both load paths are prescribed to the homogenized NTFA material law and to the full-field finite element model in order to evaluate the difference in the macroscopic stress field. A good agreement of all stress components was found, see Fig. 6.37 (oblate particles) and Fig. 6.38 (elongated particles).

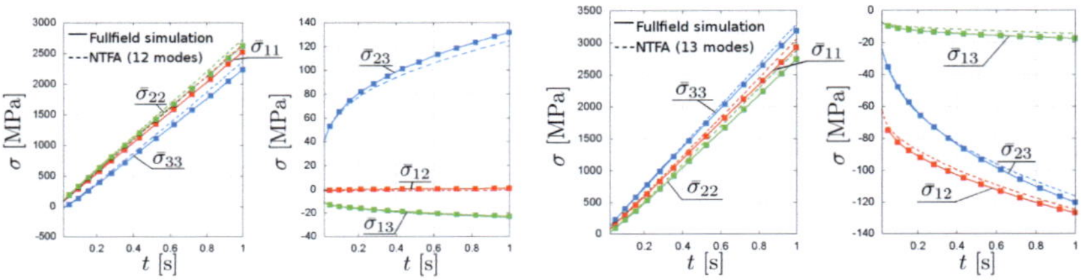

Figure 6.37: Comparison of the components of the homogenized stress tensor between full-field simulation and NTFA model for the oblate microstructure (left: $\lambda = 2$, $\dot{\bar{\varepsilon}} = \dot{\bar{\varepsilon}}_2$, 12 modes; right: $\lambda = 3$, $\dot{\bar{\varepsilon}} = \dot{\bar{\varepsilon}}_1$, 13 modes) (cf. Fritzen and Böhlke, 2011b)

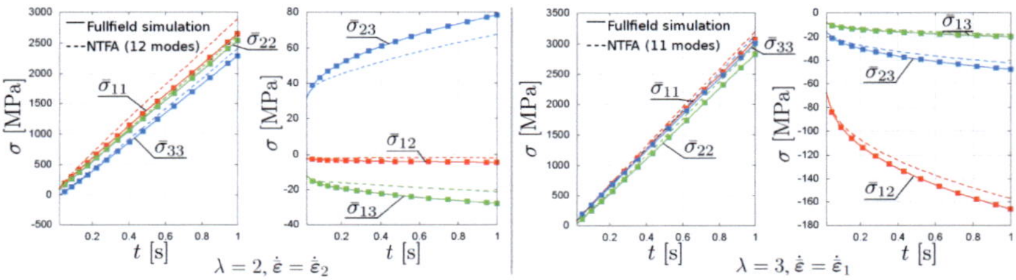

Figure 6.38: Comparison of the components of the homogenized stress tensor between full-field simulation and NTFA model for the elongated microstructure (left: $\lambda = 2$, $\dot{\bar{\varepsilon}} = \dot{\bar{\varepsilon}}_2$, 12 modes; right: $\lambda = 3$, $\dot{\bar{\varepsilon}} = \dot{\bar{\varepsilon}}_1$, 11 modes) (cf. Fritzen and Böhlke, 2011b)

Additionally, the ability of the NTFA to replicate the phase averages of the stress field is validated. Therefore, the part of the load carried by the matrix is defined via

$$\phi_{\mathrm{m}} = \frac{c_{\mathrm{m}} \langle \boldsymbol{\sigma} \rangle_{\Omega_{\mathrm{m}}} \cdot \bar{\boldsymbol{\sigma}}}{\bar{\boldsymbol{\sigma}} \cdot \bar{\boldsymbol{\sigma}}}. \tag{6.33}$$

The load fraction ϕ_{m} carried by the matrix is evaluated for each of the three different loadings

$$\dot{\bar{\varepsilon}}_3 = (2\boldsymbol{e}_1 \otimes \boldsymbol{e}_1 - \boldsymbol{e}_2 \otimes \boldsymbol{e}_2 - \boldsymbol{e}_3 \otimes \boldsymbol{e}_3) 0.015\,\mathrm{s}^{-1}, \tag{6.34}$$

$$\dot{\bar{\varepsilon}}_4 = (\boldsymbol{e}_1 \otimes \boldsymbol{e}_2 + \boldsymbol{e}_2 \otimes \boldsymbol{e}_1) 0.015\,\mathrm{s}^{-1}, \tag{6.35}$$

$$\dot{\bar{\varepsilon}}_5 = (\boldsymbol{e}_1 \otimes \boldsymbol{e}_3 + \boldsymbol{e}_3 \otimes \boldsymbol{e}_1) 0.015\,\mathrm{s}^{-1}, \tag{6.36}$$

and for each of the seven different microstructures. The time history of the load parameter is plotted versus the loading time t in Fig. 6.39 (oblate particles) and in Fig. 6.40 (elongated particles).

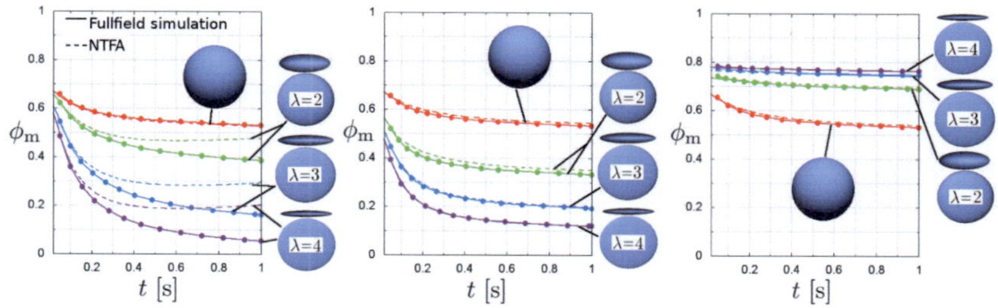

Figure 6.39: Comparison of the load fraction ϕ_{m} of the metal matrix for the macroscopic loadings $\dot{\bar{\varepsilon}}_3$ (left) $\dot{\bar{\varepsilon}}_4$ (middle) and $\dot{\bar{\varepsilon}}_5$ (right) for the oblate microstructures; for comparison the values of the isotropic structure are also shown (red curves) (cf. Fritzen and Böhlke, 2011b)

The considered loadings represent isochoric tension in the direction of the elongation/compression axis, shear in the plane spanned by the particle principle axis and the y-direction and shear in the transverse plane (y-z-plane). The results state a good qualitative agreement between the NTFA model and the full-field simulation. However, a quantitative discrepancy was observed for the isochoric tension.

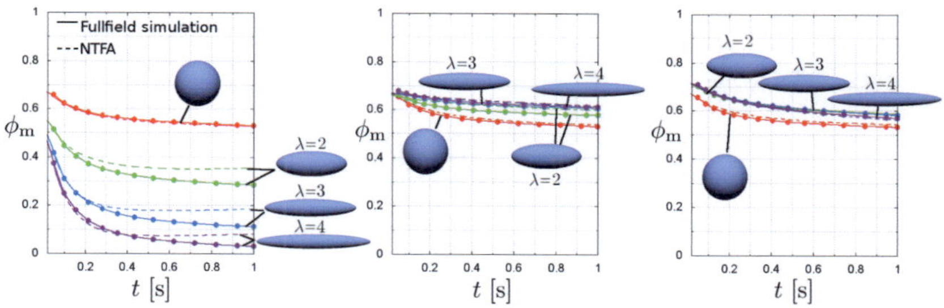

Figure 6.40: Comparison of the load fraction ϕ_{m} of the metal matrix for the macroscopic loadings $\dot{\bar{\varepsilon}}_3$ (left) $\dot{\bar{\varepsilon}}_4$ (middle) and $\dot{\bar{\varepsilon}}_5$ (right) for the elongated microstructures; for comparison the values of the isotropic structure are also shown (red curves) (cf. Fritzen and Böhlke, 2011b)

Elastic properties

The elastic properties of the examined composites are also analyzed. Therefore, the effective elastic stiffness tensor $\bar{\mathbb{C}}$ of the periodic unit cell is computed and the directional Young's modulus is investigated. The latter is defined for any normalized director $\boldsymbol{d}$ by

$$\bar{E}(\boldsymbol{d}) = \frac{1}{(\boldsymbol{d} \otimes \boldsymbol{d}) \cdot \bar{\mathbb{C}}^{-1}[\boldsymbol{d} \otimes \boldsymbol{d}]}. \tag{6.37}$$

The result is shown for the examined oblate and elongated microstructures in the x-y-plane and the y-z-plane. Note that x is the direction in which the microstructures are either stretched or compressed. Hence, the y-z-plane is the transversal plane with respect to this preferred direction. The results are shown in Fig. 6.41 (oblate particles) and 6.42 (elongated particles).

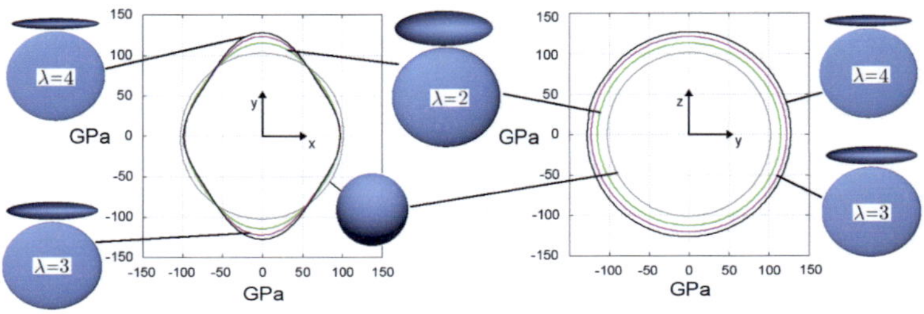

Figure 6.41: Representation of the directional Young's modulus $\bar{E}(\boldsymbol{d})$ in the x-y plane (left) and in the y-z-plane (transverse plane; right) for oblate particle reinforced metal matrix composites for $\lambda = 2, 3, 4$ compared to the Young's modulus of the isotropic microstructure (gray circle) (cf. Fritzen and Böhlke, 2011b)

Due to the anisotropic morphology of the reinforcement a significant amount of elastic anisotropy is found for all particle shapes. For both types of anisotropy the elastic modulus $E(\boldsymbol{d})$ is isotropic within the transversal plane. This implies that the microstructure consists of a sufficient number of particles.

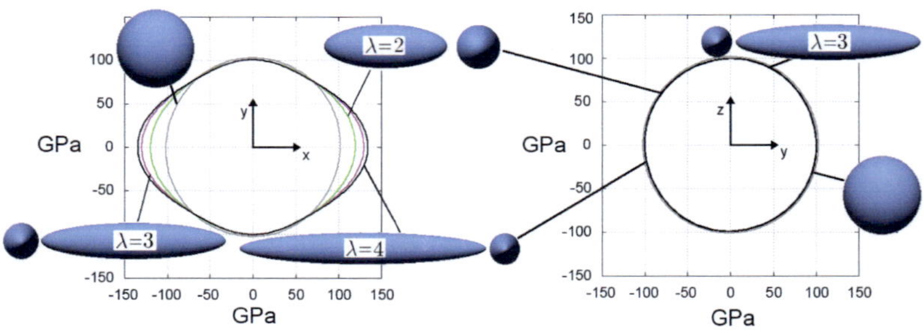

Figure 6.42: Representation of the directional Young's modulus $\bar{E}(\boldsymbol{d})$ in the x-y plane (left) and in the y-z-plane (transverse direction; right) for elongated particle reinforced metal matrix composites for $\lambda = 2, 3, 4$ compared to the Young's modulus of the isotropic microstructure (gray circle) (cf. Fritzen and Böhlke, 2011b)

6.5.3 Assessment of different mode identification strategies

A fundamental ingredient in the formulation of the NTFA is the basis of inelastic modes. More precisely the prediction of the macroscopic constitutive response of the reduced model depends on the used mode identification procedure and the numerical tests used therein. In order to examine the robustness of the method with respect to the used mode identification procedure, two of the three different strategies introduced in section 5.5 are examined in the following:

[I1] The Karhunen-Loève decomposition proposed by Roussette et al. (2009) and applied by Fritzen and Böhlke (2011b).

[I2] The Karhunen-Loève decomposition (as in **[I1]**) with subsequent orthogonalization with respect to the matrix $\hat{D}^0$.

The examined material in this test is a particulate MMC with 40 % particle volume based on a periodic Voronoi tessellation based on 6 Voronoi generators (see also section 3.3.2). The elevated particle volume fraction of 40% was chosen to examine the efficiency of the NTFA with respect to materials with very heterogeneous microstructures. The periodic discretization of the microstructure consists of a total of 48'006 nodes and 34'025 second order tetrahedral elements. As before, the particles are assumed elastic with the properties of SiC and the matrix material is assumed elasto-plastic with the inelastic material properties described earlier. The discretization of the particles and of the entire unit cell including the matrix material is shown in Fig. 6.43.

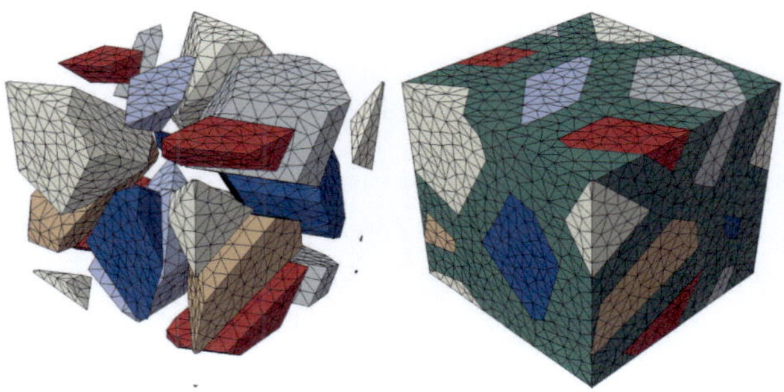

Figure 6.43: Periodic microstructure of a metal ceramic composite with 40 vol.% particles

In addition to the two different mode identification strategies [I1] and [I2], two different sets of numerical experiments were used as input data for the identification process:

[N1] Strain-driven boundary conditions defined the five orthonormal deviatoric loadings (see appendix A)

$$\bar{\boldsymbol{\varepsilon}}^{(\alpha)}(t) = \varepsilon_0(t)\boldsymbol{B}'^{(\alpha)} \quad (\alpha = 1, \dots, 5). \tag{6.38}$$

[N2] Stress-driven computations with anti-periodic traction fluctuation boundary conditions with the prescribed load

$$\bar{\boldsymbol{\sigma}}^{(\alpha)}(t) = \sigma(t)\boldsymbol{B}^{(\alpha)} \quad (\alpha = 1, \dots, 6), \tag{6.39}$$

where $\boldsymbol{B}^{(\alpha)}$ denotes an orthonormal basis of symmetric tensors defined in appendix A.

For the strain-driven computations the norm of the macroscopic strain loading was set to 5%. The stress-driven computations the load was scaled to a macroscopic von Mises equivalent stress of 350 MPa. During the numerical experiments it was found that the small number of particles leads to a non-negligible anisotropy of the mechanical response of the material.

For both strategies [I1] and [I2] a Karhunen-Loève based ansatz as proposed by Roussette et al. (2009) was used. A threshold of $\delta = 10^{-5}$ was chosen to select the eigenvalues of the matrix $\hat{K}_{\mathrm{p}}$ considered in the construction of the modes. The new mode identification approach presented in section 5.5.4 was based on the thereby identified modes. First, the mode interaction matrix $\hat{D}^0$ is determined and its real negative eigenvalues are computed. Then an orthogonal eigenbasis is identified to generate a modified set of inelastic modes such that the resulting mode interaction matrix has a block diagonal structure. In the case of one inelastic material the matrix $\hat{D}^0$ is then purely diagonal. After the linear combination of the modes, which is needed to diagonalize $\hat{D}^0$, a renormalization of the resulting plastic strain fields is indispensable. After computation of the induced stress and strain fields the interaction matrix $\hat{D}^0$ is reevaluated.

For the strain-driven numerical experiments a total of 14 modes was identified. The eigenvalues of the matrix $\hat{D}^0$ of the purely kinematic and the new thermodynamically motivated approach are compared in Tab. 6.9. While all eigenvalues are of the same order of magnitude, a quantitative variation of approximately 10% is found between the two approaches despite the same data acting as a basis for the construction.

	λ_1	λ_2	λ_3	λ_4	λ_5	λ_6	λ_7
[I1]	-28129.3	-26885.8	-25146.0	-24464.3	-23897.5	-7192.9	-6935.0
[I2]	-24411.9	-24208.3	-24351.8	-24164.0	-26584.1	-7470.1	-6837.5

	λ_8	λ_9	λ_{10}	λ_{11}	λ_{12}	λ_{13}	λ_{14}
[I1]	-6393.2	-5871.4	-3574.0	-1941.0	-1907.9	-1622.5	-1193.4
[I2]	-6578.3	-6392.5	-3619.7	-2040.2	-2079.8	-1796.8	-1409.9

Table 6.9: Comparison of the eigenvalues of the matrix $\hat{D}^0$ obtained using the kinematic mode identification procedure based on the Karhunen-Loève decomposition ([I1]) and an orthogonalization of the resulting modes with subsequent renormalization ([I2]); all values are given in MPa

The same procedure as with the strain-driven numerical experiments was performed on the inelastic strain fields computed using anti-periodic traction conditions. A total of 12 inelastic modes was found. The eigenvalues of the matrix $\hat{D}^0$ found with [I1] and after the orthogonalization step of strategy [I2] are compared in Tab. 6.10. The quantitative discrepancy between the eigenvalues found using the both methods [I1] and [I2] is slightly smaller for the stress-driven mode-identification process [N2] than for the deformation-driven process [N1] presented before. More notably, the spectrum of the modes obtained from the stress-driven calculations (procedure [N2]) decays slightly faster and is rather clustered (λ_1 to λ_5, λ_6 to λ_{10}) in comparison to the ones based on the displacement driven computations of procedure [N1].

	λ_1	λ_2	λ_3	λ_4	λ_5	λ_6
[I1]	-27655.0	-26265.5	-25451.7	-24862.9	-23378.7	-5248.1
[I2]	-25188.6	-23981.0	-26145.4	-22493.2	-23595.9	-5238.6
	λ_7	λ_8	λ_9	λ_{10}	λ_{11}	λ_{12}
[I1]	-4802.3	-4592.4	-4377.4	-4044.4	-1213.5	- 509.3
[I2]	-4677.2	-4630.9	-4501.8	-4126.0	-1213.1	-525.6

Table 6.10: Comparison of the eigenvalues of the matrix $\hat{D}^0$ obtained using the stress driven mode identification procedure based on the Karhunen-Loève decomposition ([I1]) and an orthogonalization of the resulting modes with subsequent renormalization ([I2]); all values are given in MPa

Accuracy. The computational accuracy of the different strategies was analyzed. Therefore, the time history of the macroscopic stress tensor determined from the in total eleven different loading paths are compared to predictions made by the reduced model with the different sets of modes which are all highly heterogeneous (e.g., Fig. 6.44).

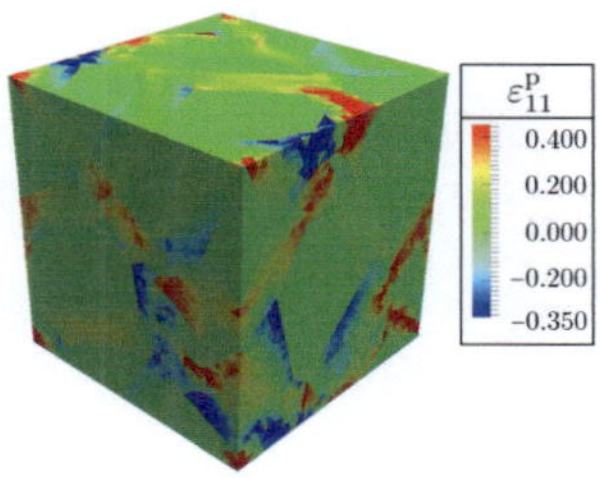

Figure 6.44: Component $\varepsilon^{\mathrm{P}}_{11}$ of the plastic strain for mode 1 ([I1], [N1])

First the five strain-driven tests are examined for the four different reduced material laws. For all five loading paths the stress components in the direction of the non-zero components of the applied load $\bar{\varepsilon}$ are compared in Fig. 6.45. Within each graph the direction prescribed macroscopic strain $\bar{\varepsilon}$ is indicated. For the first loading path parallel to $\boldsymbol{B}'^{(1)}$ the three diagonal components of the stress tensor are shown. For the load path in direction $\boldsymbol{B}'^{(2)}$ the 22- and 33- component are plotted and for the remaining loadings the stress component in the direction of the loading is depicted. Blue curves are associated with the 14 modes identified from the five strain-driven computations, whereas red curves denote the behavior of the homogenized material based on the stress-driven identification process. The arrows within the plots indicate that the strategy [I1] is displayed as a solid line while the new ansatz [I2] is shown as a dashed line. Dots denote the values associated with the finite element full-field solution.

First, the agreement between the finite element results and the homogenized material response is noteworthy. The latter is particularly important since the fields are highly heterogeneous (Fig. 6.44) due to the high particle volume fraction and, further, an anisotropic response is found. The latter can for example be seen by comparison of the final points of the first loading path when comparing the stress components $\bar{\sigma}_{22}$ and $\bar{\sigma}_{33}$. Their ratio is approximately 1.40 which is a deviation of approximately 40% from the isotropic case for which $\bar{\sigma}_{22} = \bar{\sigma}_{33}$. With respect to the basis of inelastic modes used in the different reduced computations it is found that only minor difference can be observed. These differences were more pronounced for the first two loading paths. Still the error was always smaller than 11.3%. However, most of the results were closer to the full-field simulations with errors in the range of 2-5%.

In order to compare the results of the stress-driven computations a slightly different ansatz was used. The strain history recorded in the full-field simulation was prescribed to the unit cell model. The thereby obtained results are presented in Fig. 6.46. Interestingly, all values computed using the full-field model are replicated by all four different sets of modes and for all six loadings, although the prescribed loading path was not proportional.

In summary the homogenized material response appears to be largely independent of both, the used identification strategy and the numerical testing used therein. For numerical reasons the ansatz [I2] is favorable since it simplifies many of the computational steps involved in the local Newton iteration considerably. Further, the stress-driven conditions were shown to yield slightly more accurate predictions of the mechanical response.

In all of the examples considered in this section a significant amount of non-uniformity was of the resulting fields was observed. It can be exemplified by the heterogeneity of the plastic strain field which can, e.g., be seen for the first mode ([I1], [N1]) in Fig. 6.44. Such non-uniformity highlights the importance of using non-uniform transformation strains. Notably the non-uniformity increases with the amount of different constituents and with increased amount of particles in the case of metal ceramic composites and other reinforced composites in the more general case.

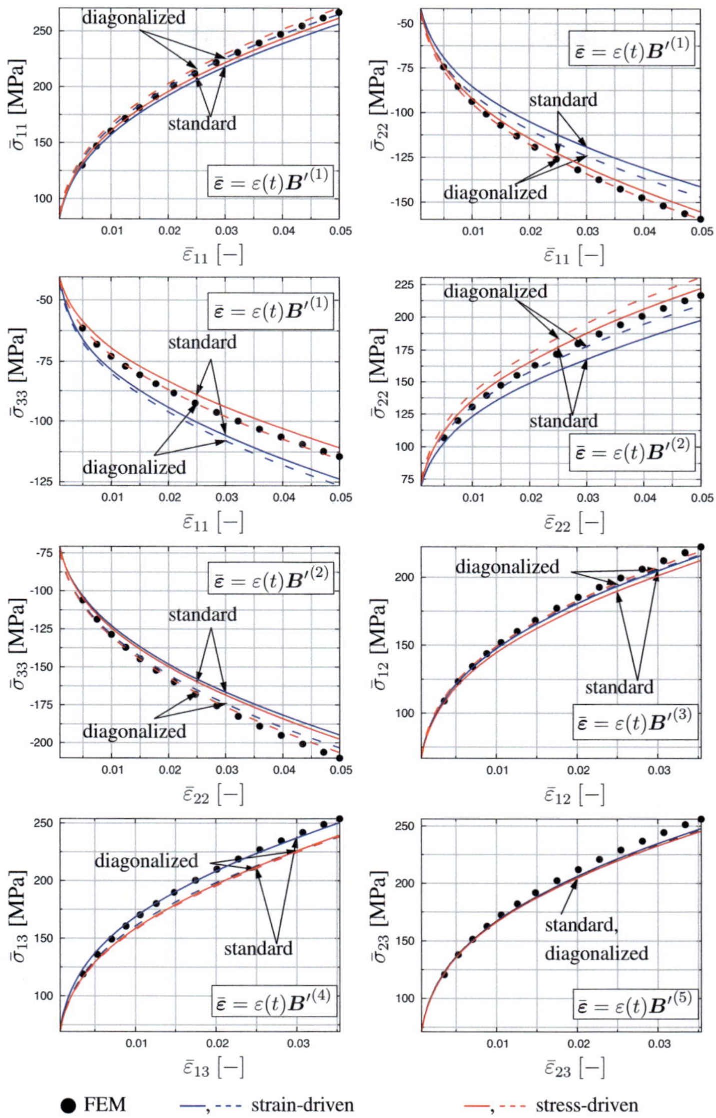

Figure 6.45: Comparison of full-field simulation (dots) and the results of the reduced model obtained from the four different mode identification methods

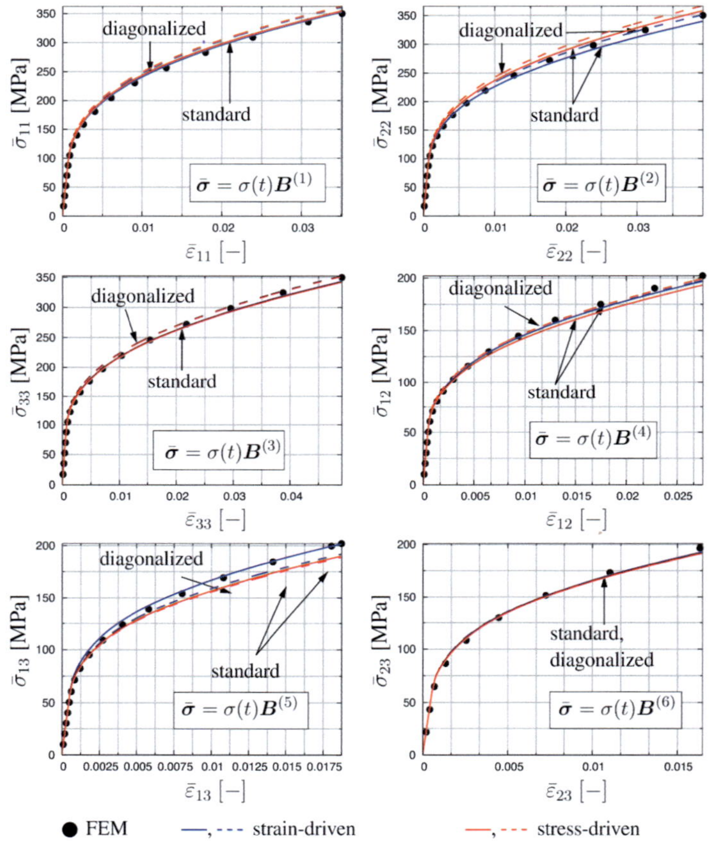

Figure 6.46: Comparison of full-field simulation (dots) and the results of the reduced model obtained from the four different mode identification methods

6.5.4 Structural applications

6.5.4.1 Tension experiment on a dog-bone specimen

The homogenized material model resulting from the NTFA applied to microstructure A is used in a structural problem. A virtual tension test of a dog-bone type tension specimen is simulated in ABAQUS/Standard. The geometric model, the spatial discretization and the boundary conditions applied to the specimen are shown in Fig. 6.47.

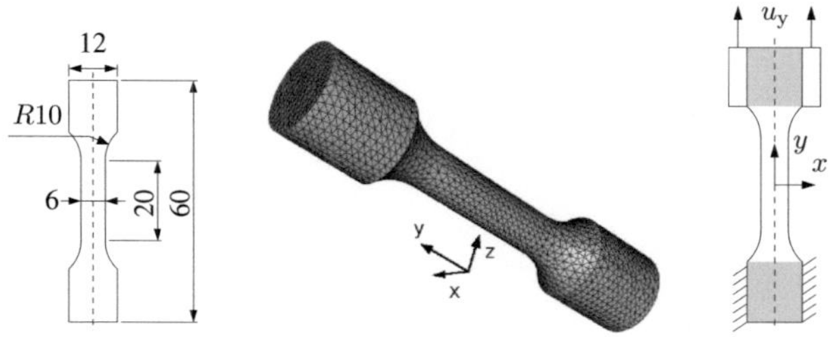

Figure 6.47: Three-dimensional tensile specimen: Geometry (unit: mm; left), spatial discretization (middle) and boundary conditions (right) (Fritzen and Böhlke, 2010b)

The force-displacement curve obtained in the numerical experiment is shown in Fig. 6.48. The main objective of this simulation is the validation of the algorithmic tangent operator. Moreover, the accuracy of the homogenized material response is assessed. Therefore, the strain history is recorded at an integration point found in the curved transition region of the test piece (see Fig. 6.49). Based on the recorded data a full-field computation is performed on the fully resolved unit cell.

The effective stress response found in the structural problem based on the NTFA model is then compared to the results of the full-field computation. The components of the macroscopic stress tensor are compared in Fig. 6.50. A considerable agreement was found for all components. In particular, the reduced model is also capable to describe the material response under multi-axial loading. Additionally, the algorithmic tangent operator was shown to lead to an excellent convergence rate of the nonlinear structural problem.

6.5.4.2 Indentation test

In order to examine the capabilities of the NTFA under pronounced discontinuities in the loading, an indentation test was modeled (see also Fritzen and Böhlke, 2010b). A semi-spherical elastic steel indenter is pushed into a bi-material. The elastic parameters of

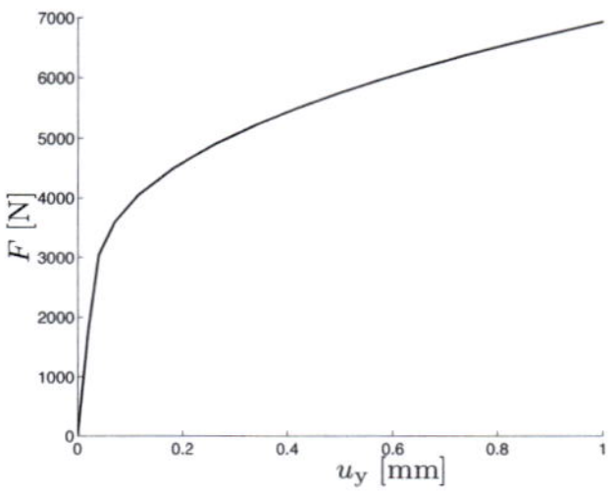

Figure 6.48: Force displacement curve (cf. Fritzen and Böhlke, 2010b)

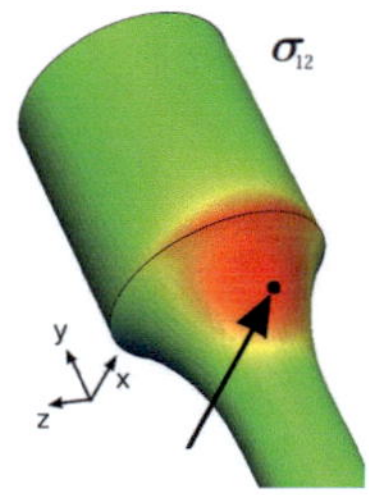

Figure 6.49: Observed integration point (cf. Fritzen and Böhlke, 2010b)

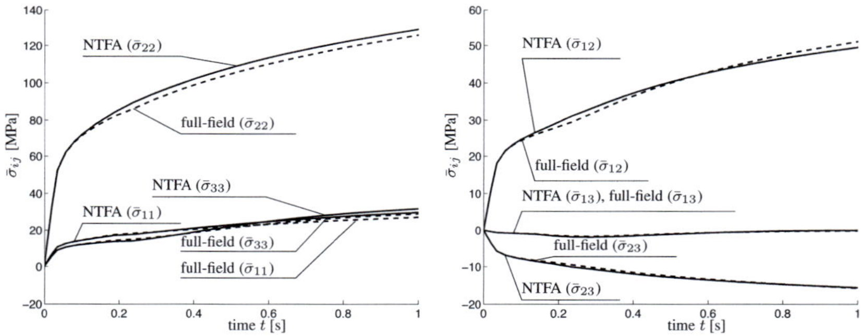

Figure 6.50: Time history of $\bar{\sigma}_{11}, \bar{\sigma}_{22}, \bar{\sigma}_{33}$ (left) and $\bar{\sigma}_{12}, \bar{\sigma}_{13}, \bar{\sigma}_{23}$ (right) for the numerical tension experiment (cf. Fritzen and Böhlke, 2010b)

steel (Young's modulus $E = 210$ GPa, Poisson ratio $\nu = 0.3$) are assumed for the indenter. The bi-material consists of a top-layer made of MMC with the homogenized material law determined for microstructure A. It covers a homogeneous base-layer made of aluminum (Fig. 6.51, lengths are given in mm). For the aluminum bulk material the same properties as for the aluminum phase in the composite layer are taken. The indenter is subjected to a total displacement of 0.1 mm equaling 0.5 % of the thickness of the MMC layer (Fig. 6.51). Hard contact conditions were assumed in normal direction and a friction coefficient of 0.1 was estimated.

At the end of the indentation process the distribution of the five mode stimulation coefficients has been captured (Fig. 6.52). A large degree of heterogeneity can be observed. Different modes are active in different regions. Particularly, multiple modes are active in the entire contact region.

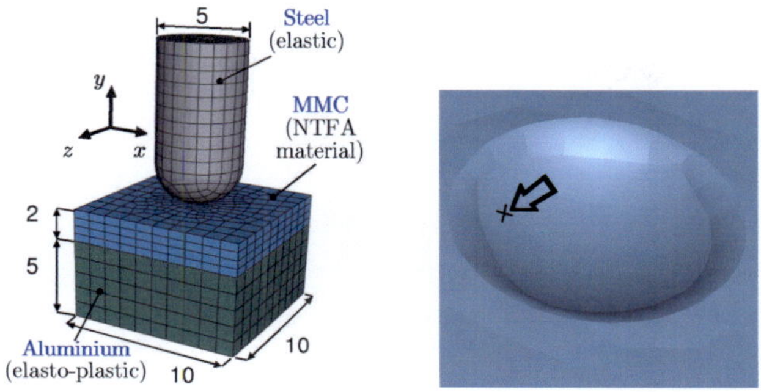

Figure 6.51: Geometry of the indentation test and integration point at which the strain path history is recorded (cf. Fritzen and Böhlke, 2010b)

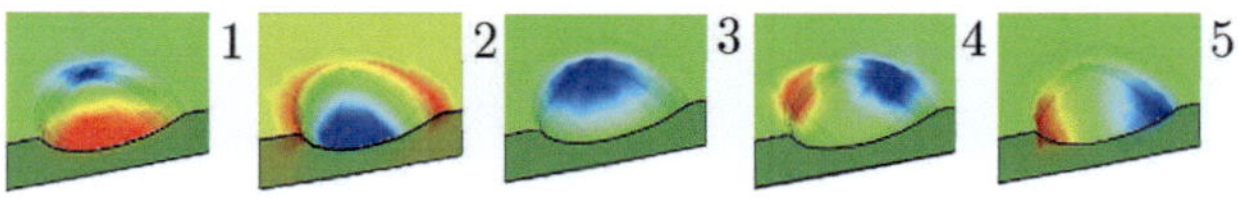

Figure 6.52: Distribution of the mode coefficients at the end of the increment (cut through MMC layer) (cf. Fritzen and Böhlke, 2010b)

Similarly to the previous example, the time history of the stress and strain tensor were recorded. The integration point data was used to run a full-field simulation on the entire unit cell. The components of the macroscopic stress tensor of the full-field simulation and the homogenized material model are plotted in Fig. 6.53 over time*. Due to the position of the integration point out of the center of the contact region, the deformation in the first few steps of the analysis was dominated by neighboring regions that have already been in contact. Then the area containing nodes directly associated with the integration point get into contact and a load reversal is observed (Fig. 6.53, e.g. $\bar{\sigma}_{22}, \bar{\sigma}_{12}, \bar{\sigma}_{13}$). Although only five modes were used in the analysis, the stress predicted by the homogenized material model shows considerable agreement to the full-field simulation.

During the simulation of the contact problem, the algorithmic stiffness operator derived allowed for good convergence rates. Notably, the homogenized model produced almost the same results as the full-field simulation while the latter required significantly more load increments in order to converge. This is due to the local deformation getting excessive. The previous observations underline the robustness of the method in practical applications.

*Due to the non-proportionality of the straining direction a plot with respect to individual strain components does not appear sensible.

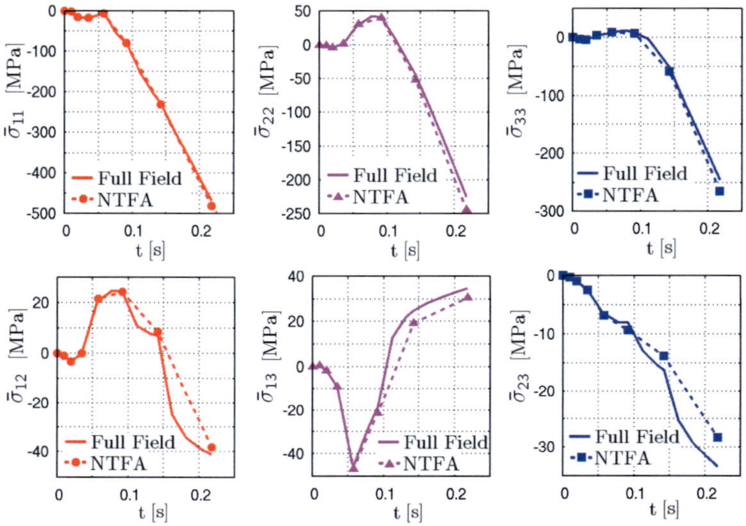

Figure 6.53: Components of the macroscopic stress: full-field simulation vs. NTFA model (cf. Fritzen and Böhlke, 2010b)

Chapter 7

Summary and conclusions

In the presented work different aspects of the computational homogenization of the thermo-mechanical properties of microheterogeneous materials are investigated. The developed methodology is applied to polycrystalline aggregates, porous metals and particle reinforced composite materials. The concept of model microstructures introduced in chapter 3 is used to approximate the complex microstructure of real materials by artificial models. The statistical distribution of the microstructure is considered based on a parametrization of the microgeometry using random variables. A particular focus in all of the presented microstructural geometries is the periodicity of the underlying topology, which has pronounced implications on the size of the unit cell required to attain statistical representativity of the results. Algorithms for the periodic spatial discretization of the examined geometries are developed. They allow for an efficient implementation of the periodicity constraints in the context of finite element simulations. In order to argue on the representativity of the model materials, different statistical measures are introduced and evaluated for different structures with success. Anisotropic modifications of some of the presented model microstructures are considered. These modifications are used to systematically study the effect of the morphological anisotropy for elasto-visco-plastic polycrystals and particle reinforced metal ceramic composites.

The thermo-mechanical multiscale problem is addressed in chapter 4. A specialization for two-scale problems is given and rules for the thermo-mechanical scale transition are proposed. Different boundary conditions are briefly discussed in order to outline capabilities of the presented methods for non-periodic unit cells, which can usually not be subjected to periodic field constraints. For the special case of linear thermal and thermo-elastic homogenization an algorithm for the computational homogenization is presented.

A brief reference to (semi-)analytical approaches is given and the problem of the intrinsic path-dependency of physically nonlinear materials is addressed. In order to homogenize the behavior of such nonlinear materials it is decided on the Nonuniform Transformation Field Analysis (NTFA, Michel and Suquet (2003, 2004); Fritzen and Böhlke (2009a,b, 2010b, 2011b)). The initial representation of the method (Michel and Suquet, 2003) is reformulated based on Fritzen and Böhlke (2010b) in chapter 5. Extensions for the consideration of thermo-mechanical problems are proposed and a detailed account for the properties of the obtained system matrices describing the homogenized material response is given. A weak point of the method is found in the approximation of the effective yield criterion following the proposals of Michel and Suquet (2003, 2004). In order to investigate the plausibility of the coupled and uncoupled model a simple test case is considered. It is shown that the uncoupled model can lead to pronounced over-predictions of the macroscopic stresses. Interestingly the theoretic error for a homogeneous model problem is found in numerical computations on heterogeneous microstructures performed by Michel and Suquet (2004). This reversal is compensated by the coupled model which can be shown to reproduce the constitutive response of the homogeneous reference material exactly. Based on the properties of the system matrices a thermo-dynamically motivated post-orthogonalization procedure is proposed for the mode identification process. In addition, a modified normalization condition is formulated which helps to improve the efficiency for strongly localizing materials.

With respect to the numerical implementation of the NTFA it is decided on the finite element method. All operators required in the course of the implementation are described in detail and three different mode identification strategies are proposed. An algorithm for the implicit time integration of the constitutive response of the homogenized material model is given for the backward Euler scheme. An analytical expression for the tangent stiffness operator is presented and the symmetry of the latter is proven.

The NTFA is applied to particle reinforced composites. A validation of the predictions of the method is performed based on different random load paths. These are applied to the fully resolved microscopic unit cell and to the homogenized material and the effective stress response of both computations is compared. An excellent agreement is found even for the case of partial load reversal and for loadings that have not been investigated in the identification process. Moreover, different particle volume fractions from 18.2% up to 40% are investigated. For all examples the deviations between the NTFA predictions and the full-field results can be considered small. The computational efficiency of the method is excellent and, moreover, it is independent of the fine scale mesh size. More precisely, the only parameter influencing the memory and CPU requirements is found in the number of modes. In all of the presented examples the latter was found in the range of five to thirteen. With respect to the mode identification process, three different techniques are examined and a detailed comparison of the two approaches involving the Karhunen-Loève decomposition showed the robustness of the method with respect to the choice of the modes.

Based on the proposed implicit time integration of the homogenized material model, user material routines were developed for both, ABAQUS and different self-written finite element implementations. The efficiency of these routines qualifies the method for the use in large scale structural computations. Examples for the latter are investigated in terms of a dog-bone style tensile specimen and an indentation test involving contact. The NTFA predictions have been validated a posteriori by prescribing the recorded strain history of selected integration points to the discretized microstructure. The results are then compared to the model predictions. It was found that even for the complex load case examined in the indentation test, the computational predictions of the NTFA were accurate in comparison to the full field results. Additionally, the algorithmic tangent modulus was found to lead to good convergence rates even for large load increments. A point which is surprising is that the NTFA in the presented proposal tends to slightly underestimate the macroscopic stresses in comparison to the full-field computations, whereas many other homogenization techniques overestimate the macroscopic stress.

Although the capabilities of the NTFA for the examined class of materials are promising, the method has several limitations defining future scientific challenges. The key problem of the method is found in the fact that the effective dissipation potential defining the evolution law for the internal variables is chosen on a phenomenological basis. While the results for the von Mises plasticity model presented in works by the author (Fritzen and Böhlke, 2009a,b, 2010b, 2011b,a) and others (Michel and Suquet, 2003, 2004; Roussette et al., 2009; Michel and Suquet, 2009), the limitation to isotropic inelastic constituents is a pronounced disadvantage.

Although the potential associated with the coupled model of Michel and Suquet (2003) is in some sense a heuristic choice, the motivations presented in the current work highlight that the method is mechanically motivated, but remains an approximation. With respect to future extensions of the method a generalization with respect to other constitutive laws is required. This involves the development of a standardized procedure based on rather weak assumptions on the dissipation potential. The author is confident that the physical meaning of the system matrices derived in chapter 5 and the bounds derived for individual components of these matrices will be of use in these generalizations.

With respect to large strain applications the method is not suitable in its present form. However, the results obtained using the NTFA have revealed the capabilities of order reduction methods (e.g., Ryckelynck and Benziane, 2010; Ladevèze et al., 2010) in the context of computer based homogenization. Surprisingly the scientific community has so far mostly neglected these methods for multiscale problems. The NTFA has shown that based on a micro-mechanical concept the efficiency of the order reduction methods can be significantly improved with respect to computational time and memory requirements, while a sufficient accuracy of the macroscopic response is preserved.

Its role model character has motivated new developments in the linkage of micro-mechanics and order reduction techniques. A combination of the capabilities of more general approaches such as the hyperreduction method (Ryckelynck, 2009; Ryckelynck et al., 2010; Ryckelynck and Benziane, 2010) with the advantages of the NTFA derived from the micromechanical motivation are subject of ongoing research. This ongoing work raises hope that the generalization to new constitutive models and possibly large strain formulations is only a matter of time.

Another approach to computational homogenization was pursued in terms of a large scale Monte Carlo type study for porous metals. The aim of this study was the systematic comparison of many of the existing (semi-)analytical constitutive laws to the response obtained from high-resolution finite element computations involving periodicity constraints. Based on the computational data a modification of the popular Gurson-Tvergaard-Needleman model was identified involving only a single additional parameter. The new model can replicate the pressure dependency of the yield stress of all of the more than 500 computations at pore volume fractions ranging from 0.1 to 30% to an excellent extent. Thereby, the capabilities of the identification of both the mathematical structure and the numerical coefficients of (semi-)phenomenological material models based on numerical experiments are highlighted.

Appendix A

Notation

A.1 General tensorial notation

Throughout the presented work an index free notation for tensors is adopted. First-order tensors are denoted by lower case letters in bold type-setting, e.g. $\boldsymbol{u}, \boldsymbol{n}$. Second-order tensors are referred to by bold upper case letters and bold greek symbols, e.g. $\boldsymbol{B}, \boldsymbol{\sigma}$. For fourth-order tensors are denoted by $\mathbb{A}, \mathbb{C}$ and alike. No summation convention is used in this work. Each of tensor can also be written with respect to a given orthonormal basis $\boldsymbol{e}_i$ ($i = 1, 2, 3$) according to

$$\boldsymbol{a} = \sum_{i=1}^{3} a_i \boldsymbol{e}_i, \tag{A.1}$$

$$\boldsymbol{A} = \sum_{i=1}^{3} \sum_{j=1}^{3} A_{ij} \boldsymbol{e}_i \otimes \boldsymbol{e}_j, \tag{A.2}$$

$$\mathbb{A} = \sum_{i=1}^{3} \sum_{j=1}^{3} \sum_{k=1}^{3} \sum_{l=1}^{3} A_{ijkl} \boldsymbol{e}_i \otimes \boldsymbol{e}_j \otimes \boldsymbol{e}_k \otimes \boldsymbol{e}_l. \tag{A.3}$$

The transposition of second-order tensors in the assumed orthonormal basis is defined by a permutation of the coefficients A_{ij}

$$\boldsymbol{A}^{\mathsf{T}} = \sum_{i=1}^{3} \sum_{j=1}^{3} A_{ji} \boldsymbol{e}_i \otimes \boldsymbol{e}_j. \tag{A.4}$$

A tensor satisfying the relation $\boldsymbol{A}^{\mathsf{T}} = \boldsymbol{A}$ is called symmetric. Any tensor $\boldsymbol{A}$ can be additively decomposed by a symmetric $\boldsymbol{A}_{\text{sym}}$ and skew part $\boldsymbol{A}_{\text{skw}}$ according to

$$\boldsymbol{A} = \boldsymbol{A}_{\text{sym}} + \boldsymbol{A}_{\text{skw}}. \tag{A.5}$$

For the skew part of $\boldsymbol{A}$ the relation

$$\boldsymbol{A}_{\text{skw}} = -\boldsymbol{A}_{\text{skw}}^{\mathsf{T}} \tag{A.6}$$

holds. The symmetric part can further be separated into a deviatoric tensor $\boldsymbol{A}'$ and a spherical tensor $\boldsymbol{A}^{\circ}$

$$\boldsymbol{A}_{\text{sym}} = \boldsymbol{A}' + \boldsymbol{A}^{\circ}, \qquad \boldsymbol{A}' = \boldsymbol{A}_{\text{sym}} - \boldsymbol{A}^{\circ}, \qquad \boldsymbol{A}^{\circ} = \frac{1}{3}\operatorname{tr}(\boldsymbol{A})\boldsymbol{I}. \tag{A.7}$$

Similarly to the transposition of second-order tensors, the major transposition of fourth-order tensors is defined by

$$\mathbb{A}^{\mathsf{T}} = \sum_{i=1}^{3}\sum_{j=1}^{3}\sum_{k=1}^{3}\sum_{l=1}^{3} A_{klij}\boldsymbol{e}_i \otimes \boldsymbol{e}_j \otimes \boldsymbol{e}_k \otimes \boldsymbol{e}_l. \tag{A.8}$$

A tensor $\mathbb{C}$ has both minor symmetries, if

$$C_{ijkl} = C_{ijlk} = C_{jikl} \quad (i,j,k,l = 1,2,3) \tag{A.9}$$

holds. With the Kronecker symbol

$$\delta_{ij} = \begin{cases} 1 & i = j \\ 0 & i \neq j \end{cases}, \tag{A.10}$$

the identity on first-order tensors $\boldsymbol{I}$ and on second-order tensors $\mathbb{I}$ are expressed by

$$\boldsymbol{I} = \sum_{i=1}^{3}\sum_{j=1}^{3} \delta_{ij}\boldsymbol{e}_i \otimes \boldsymbol{e}_j, \tag{A.11}$$

$$\mathbb{I} = \sum_{i=1}^{3}\sum_{j=1}^{3}\sum_{k=1}^{3}\sum_{l=1}^{3} \delta_{ik}\delta_{jl}\boldsymbol{e}_i \otimes \boldsymbol{e}_j \otimes \boldsymbol{e}_k \otimes \boldsymbol{e}_l. \tag{A.12}$$

The identity on symmetric second-order tensors $\mathbb{I}^{\text{s}}$ is given by

$$\mathbb{I}^{\text{s}} = \sum_{i=1}^{3}\sum_{j=1}^{3}\sum_{k=1}^{3}\sum_{l=1}^{3} \frac{1}{2}(\delta_{ik}\delta_{jl} + \delta_{il}\delta_{jk})\boldsymbol{e}_i \otimes \boldsymbol{e}_j \otimes \boldsymbol{e}_k \otimes \boldsymbol{e}_l. \tag{A.13}$$

Due to its importance the operators $\mathbb{P}_1^{\text{iso}}$ and $\mathbb{P}_2^{\text{iso}}$ acting as identities on the space of spherical and deviatoric tensors are also defined

$$\mathbb{P}_1^{\text{iso}} = \frac{1}{3}\boldsymbol{I} \otimes \boldsymbol{I}, \qquad \mathbb{P}_2^{\text{iso}} = \mathbb{I}^{\text{s}} - \mathbb{P}_1^{\text{iso}}. \tag{A.14}$$

They are commonly referred to as isotropic projectors. The standard inner products on second and fourth-order tensors are defined by the trace operator $\mathrm{tr}(\bullet)$ according to

$$\boldsymbol{A}\cdot\boldsymbol{B}=\mathrm{tr}(\boldsymbol{A}^{\mathsf{T}}\boldsymbol{B}),\qquad \mathbb{A}\cdot\mathbb{B}=\mathrm{tr}(\mathbb{A}^{\mathsf{T}}\mathbb{B}). \tag{A.15}$$

These inner products induce the norms

$$\|\boldsymbol{A}\|_2=\sqrt{\boldsymbol{A}\cdot\boldsymbol{A}},\qquad \|\mathbb{A}\|_2=\sqrt{\mathbb{A}\cdot\mathbb{A}}. \tag{A.16}$$

The following linear operations on tensors are used without further precision

$$\boldsymbol{A}\boldsymbol{b}=\sum_{i=1}^{3}\sum_{j=1}^{3}A_{ij}b_j\boldsymbol{e}_i, \tag{A.17}$$

$$\mathbb{C}[\boldsymbol{A}]=\sum_{i=1}^{3}\sum_{j=1}^{3}\sum_{k=1}^{3}\sum_{l=1}^{3}C_{ijkl}A_{kl}\boldsymbol{e}_i\otimes\boldsymbol{e}_j, \tag{A.18}$$

$$\mathbb{A}\mathbb{C}=\sum_{i=1}^{3}\sum_{j=1}^{3}\sum_{k=1}^{3}\sum_{l=1}^{3}\sum_{m=1}^{3}\sum_{n=1}^{3}A_{ijkl}C_{klmn}\boldsymbol{e}_i\otimes\boldsymbol{e}_j\otimes\boldsymbol{e}_m\otimes\boldsymbol{e}_n. \tag{A.19}$$

For tensors $\boldsymbol{A}$, $\mathbb{C}$ denoting bijective operations on first and second-order tensors respectively, unique inverse elements $\boldsymbol{A}^{-1}, \mathbb{C}^{-1}$ satisfying

$$\boldsymbol{A}^{-1}\boldsymbol{A}=\boldsymbol{I},\qquad \mathbb{C}^{-1}\mathbb{C}=\mathbb{I}. \tag{A.20}$$

Note that according to the previous definition the tensor $\mathbb{I}^{\mathrm{s}}$ is not invertible. Therefore, the inverse of symmetric fourth-order tensors of rank six is defined for convenience according to

$$\mathbb{C}^{-1}\mathbb{C}=\mathbb{I}^{\mathrm{s}}, \tag{A.21}$$

where $\mathbb{C}^{-1}$ and $\mathbb{C}$ are both symmetric.

A.2 Special tensorial basis

Symmetric second-order tensors are of major significance in this monograph. Therefore, an orthonormal basis for these tensors can be defined by (similarly to, e.g., Federov, 1968)

$$\left.\begin{aligned}
\boldsymbol{B}^{(1)} &= \boldsymbol{e}_1\otimes\boldsymbol{e}_1\\
\boldsymbol{B}^{(2)} &= \boldsymbol{e}_2\otimes\boldsymbol{e}_2\\
\boldsymbol{B}^{(3)} &= \boldsymbol{e}_3\otimes\boldsymbol{e}_3\\
\boldsymbol{B}^{(4)} &= \tfrac{\sqrt{2}}{2}(\boldsymbol{e}_1\otimes\boldsymbol{e}_2+\boldsymbol{e}_2\otimes\boldsymbol{e}_1)\\
\boldsymbol{B}^{(5)} &= \tfrac{\sqrt{2}}{2}(\boldsymbol{e}_1\otimes\boldsymbol{e}_3+\boldsymbol{e}_3\otimes\boldsymbol{e}_1)\\
\boldsymbol{B}^{(6)} &= \tfrac{\sqrt{2}}{2}(\boldsymbol{e}_2\otimes\boldsymbol{e}_3+\boldsymbol{e}_3\otimes\boldsymbol{e}_2)
\end{aligned}\right\} \tag{A.22}$$

For convenience an orthonormal basis of the five-dimensional space of deviatoric tensors is also defined via

$$\left.\begin{aligned} \boldsymbol{B}'^{(1)} &= \tfrac{\sqrt{6}}{3}\left(\boldsymbol{e}_1 \otimes \boldsymbol{e}_1 - \tfrac{1}{2}\left(\boldsymbol{e}_2 \otimes \boldsymbol{e}_2 + \boldsymbol{e}_3 \otimes \boldsymbol{e}_3\right)\right) \\ \boldsymbol{B}'^{(2)} &= \tfrac{\sqrt{2}}{2}\left(\boldsymbol{e}_2 \otimes \boldsymbol{e}_2 - \boldsymbol{e}_3 \otimes \boldsymbol{e}_3\right) \\ \boldsymbol{B}'^{(3)} &= \tfrac{\sqrt{2}}{2}\left(\boldsymbol{e}_1 \otimes \boldsymbol{e}_2 + \boldsymbol{e}_2 \otimes \boldsymbol{e}_1\right) \\ \boldsymbol{B}'^{(4)} &= \tfrac{\sqrt{2}}{2}\left(\boldsymbol{e}_1 \otimes \boldsymbol{e}_3 + \boldsymbol{e}_3 \otimes \boldsymbol{e}_1\right) \\ \boldsymbol{B}'^{(5)} &= \tfrac{\sqrt{2}}{2}\left(\boldsymbol{e}_2 \otimes \boldsymbol{e}_3 + \boldsymbol{e}_3 \otimes \boldsymbol{e}_2\right) \end{aligned}\right\} \tag{A.23}$$

A.3 Vector-matrix representation of symmetric tensors

By virtue to the symmetric basis (A.22), any symmetric tensor $\boldsymbol{A}$ can alternatively be expressed by a six-dimensional vector $\hat{A}$ with column components

$$A_\alpha = \boldsymbol{A} \cdot \boldsymbol{B}^{(\alpha)} \quad (\alpha = 1, \ldots, 6). \tag{A.24}$$

Given two symmetric tensors $\boldsymbol{A}$, $\boldsymbol{D}$, the following identities hold:

$$\|\boldsymbol{A}\|_2 = \|\hat{A}\|_2 = \sqrt{\sum_{\alpha=1}^{6} A_\alpha^2}, \tag{A.25}$$

$$\boldsymbol{A} \cdot \boldsymbol{D} = \hat{A}^\mathsf{T} \hat{D} = \hat{A} \cdot \hat{D} = \sum_{\alpha=1}^{6} A_\alpha D_\alpha. \tag{A.26}$$

Further, any fourth-order tensor $\mathbb{C}$ having both minor symmetries can be reduced to a six by six matrix notation

$$(\hat{C})_{\alpha\beta} = \mathbb{C} \cdot (\boldsymbol{B}^{(\alpha)} \otimes \boldsymbol{B}^{(\beta)}) = (\hat{C})_{\beta\alpha} \quad (\alpha, \beta = 1, \ldots, 6). \tag{A.27}$$

If $\mathbb{C}$ is also major symmetric ($\mathbb{C}^\mathsf{T} = \mathbb{C}$), then the matrix $\hat{C}$ is symmetric, too. The identity operator $\mathbb{I}^\mathrm{s}$ can then be expressed as the six-dimensional identity matrix. Similarly the projectors $\mathbb{P}_1^\mathrm{iso}$ and $\mathbb{P}_2^\mathrm{iso}$ are

$$\hat{P}_1^\mathrm{iso} = \frac{1}{3}\begin{pmatrix} 1 & 1 & 1 & 0 & 0 & 0 \\ 1 & 1 & 1 & 0 & 0 & 0 \\ 1 & 1 & 1 & 0 & 0 & 0 \\ 0 & 0 & 0 & 0 & 0 & 0 \\ 0 & 0 & 0 & 0 & 0 & 0 \\ 0 & 0 & 0 & 0 & 0 & 0 \end{pmatrix}, \quad \hat{P}_2^\mathrm{iso} = \frac{1}{3}\begin{pmatrix} 2 & -1 & -1 & 0 & 0 & 0 \\ -1 & 2 & -1 & 0 & 0 & 0 \\ -1 & -1 & 2 & 0 & 0 & 0 \\ 0 & 0 & 0 & 3 & 0 & 0 \\ 0 & 0 & 0 & 0 & 3 & 0 \\ 0 & 0 & 0 & 0 & 0 & 3 \end{pmatrix}. \tag{A.28}$$

Using the presented notation the tensor operation $\mathbb{C}[\boldsymbol{A}]$ has an equivalent matrix vector notation

$$\boldsymbol{B} = \mathbb{C}[\boldsymbol{A}] \quad \Leftrightarrow \quad \hat{B} = \hat{C}\hat{A}. \tag{A.29}$$

Given symmetric tensors $\boldsymbol{A}, \boldsymbol{B}$ the dyadic product in the presented matrix notation is given by

$$\mathbb{C} = \boldsymbol{A} \otimes \boldsymbol{B} \quad \Leftrightarrow \quad \hat{C} = \hat{A}\hat{B}^{\mathsf{T}}. \tag{A.30}$$

The restricted inverse of a minor symmetric fourth-order tensor $\mathbb{C}$ is

$$\mathbb{S} = \mathbb{C}^{-1} \quad \Leftrightarrow \quad \hat{S} = \hat{C}^{-1}. \tag{A.31}$$

Finally, the composition of two fourth-order tensors is equivalent to the matrix product

$$\mathbb{C} = \mathbb{A}\mathbb{B} \quad \Leftrightarrow \quad \hat{C} = \hat{A}\hat{B}. \tag{A.32}$$

A.4 List of abbreviations

FEM	finite element method
CPU	central processing unit
NTFA	nonuniform transformation field analysis
TFA	transformation field analysis
POD	proper orthogonal decomposition
PGD	proper generalized decomposition
GSM	generalized standard material
FFT	Fast Fourier Transformation
CODF	crystallite orientation distribution function
UC	unit cell
VC	Voronoi cell
SV	standard Voronoi tessellation
HC	hard-core Voronoi tessellation
CV	centroidal Voronoi tessellation
MMC	metal matrix composite
GTN model	Gurson-Tvergard-Needleman model
LA	limit analysis
UKBC	uniform kinematic boundary conditions
PKBC	periodic fluctuation kinematic boundary conditions
MCK model	Monchiet-Charkaluk-Kondo model
S&W model	Sun-Wang model
Gurson-HS model	Gurson-Hashin Shtrikman model

Index

analysis of the system matrices, 78
analytical methods, 61
anisotropic grain morphology, 32

balance of mass, 13
Boltzmann continuum, 13
boundary conditions, 14

Cauchy continuum, 13
Cauchy stress tensor, 13
Cauchy's lemma, 13
cavity strain, 53
centroidal Voronoi tessellation, 30
Clausius-Duhem inequality, 15
coupled model (modified), 86
coupled NTFA model, 82

deformation gradient, 11
displacement gradient, 11
dissipation inequality, 16
dissipation potential, 17
divergence, 10
dual dissipation potential, 17

effective displacement gradient, 53
effective strain tensor, 53
effective stress tensor, 53
entropy, 14
evolution of mode activity, 80

finite element operators, 97
first law of thermodynamics, 14
force field, 12
Fourier heat conduction, 17
free energy, 15

Generalized Standard Materials (GSM), 17
gradient, 10

hardcore condition, 31
heat flux, 14
heating, 14
Helmholz free energy, 15

implicit time integration, 99
inelastic modes, 74
infinitesimal rotation, 12
infinitesimal strain tensor, 12
internal energy, 14
internal variables, 14
isotropic hardening potential, 75

k-nearest neighbor distance, 25
Karhunen-Loève decomposition, 89
kinematic hardening metric, 79
kinematic hardening potential, 75

linear momentum, 14

mass density, 12
material time derivative, 10
maximum Feret, 25
mesh generation (crystal aggregates), 34
mesh generation (particulate materials), 42
mesh generation (porous materials), 49
metal matrix composite, 41
minimum Feret, 25
mode activity coefficients, 74
mode identification, 87
mode interaction matrix, 79
model microstructures, 21
modified coupled model, 86
motion, 9
multiscale problem, 51

n-point auto-correlation function, 23
nonuniform transformation field analysis, 71

numerical implementation (NTFA), 95

order reduction methods, 72

particle reinforced composites, 41
periodic spatial discretization, 34
periodicity constraint, 29
porous materials, 46

second law of thermodynamics, 15
semi-analytical methods, 61
specific strain energy, 75
statistical ergodicity, 22
statistical homogeneity, 22
statistical isotropy, 24

temperature, 14
thermal energy, 75
thermodynamic force, 16
thermodynamical mode identification, 92
traction vector, 13
two-scale problem, 51

uncoupled NTFA model, 82

von Mises plasticity, 18
Voronoi generators, 29
Voronoi tessellation, 29

Bibliography

Adams, B., Olson, T., 1998. The mesostructure - properties linkage in polycrystals. Progress in Material Science 43, 1–88.

Armstrong, P., Frederick, C., 1966. A mathematical representation of the multiaxial Bauschinger effect. Tech. Rep. RD/B/N731, Berkeley Nuclear Laboratories, Berkeley, U.K.

Aurenhammer, F., 1991. Voronoi Diagrams - A Survey of a Fundamental Geometric Data Structure. ACM Computing Surveys 23 (3), 345–405.

Ball, J., 1977. Convexity Conditions and Existence Theorems in Nonlinear Elasticity. Archive of Rational Mechanics and Analysis 63 (4), 337–403.

Barbe, F., Decker, L., Jeulin, D., Cailletaud, G., 2001. Intergranular and intragranular behavior of polycrystalline aggregates. Part 1: F.E. Model. International Journal of Plasticity 17, 513–536.

Barber, C., Dobkin, D., Huhdanpaa, H., 1996. The Quickhull algorithm for convex hulls. ACM Trans. Math. Softw. 22 (4), 469–483.

Bathe, K., 2002. Finite-Elemente-Methoden, 2nd Edition. Springer, Berlin.

Benveniste, Y., Dvorak, G. J., Chen, T., 1991. On diagonal and elastic symmetry of the approximate effective stiffness tensor of heterogeneous media. Journal of the Mechanics and Physics of Solids 39 (7), 927 – 946.

Bertram, A., Böhlke, T., 1999. Simulation of texture induced elastic anisotropy of polycrystalline copper. Computational Materials Science 16 (1-4), 2 – 9.

Besson, J. (Ed.), 2004. Local approach to fracture. Ecole des Mines de Paris–Les Presses, Paris.

Besson, J., Cailletaud, G., Chaboche, J.-L., Forest, S., 2010. Nonlinear Mechanics of Materials. Springer Science+Business Media B.V, Dordrecht.

Bilger, N., Auslender, F., Bornert, M., Michel, J.-C., Moulinec, H., Suquet, P., Zaoui, A., 2005. Effect of a nonuniform distribution of voids on the plastic response of voided materials: a computational and statistical analysis. International Journal of Solids and Structures 42 (2), 517 – 538.

Bollhoefer, M., Saad, Y., 2006. Multilevel preconditioners constructed from inverse-based ILUs. SIAM J. Sci. Comput., Special Issue on the 8-th Copper Mountain Conference 5 (27).

Böhlke, T., 2005. Application of the maximum entropy method in texture analysis. Computational Materials Science 32 (3-4), 276 – 283, IWCMM.

Böhlke, T., Bertram, A., 2001. The evolution of Hooke's law due to texture development in FCC polycrystals. International Journal of Solids and Structures 38 (52), 9437 – 9459.

Böhlke, T., Fritzen, F., Jöchen, K., Tsotsova, R., 2009. Numerical methods for the quantification of the mechanical properties of crystal aggregates with morphologic and crystallographic texture. International Journal for Material Forming 2, 915–917.

Carrere, N., Maire, J. F., Kruch, S., Chaboche, J. L., 2004. Multiscale analysis of SiC/Ti composites. Materials Science and Engineering A 365 (1-2), 275 – 281, Multiscale Materials Modelling.

Chaboche, J.-L., 1991. On some modifications of kinematic hardening to improve the description of ratcheting effects. International Journal of Plasticity (7), 661–678.

Chawla, N., Chawla, K., 2006. Metal matrix composites. Springer, New York, NY.

Chawla, N., Deng, X., Schnell, D., 2006. Thermal expansion anisotropy in extruded SiC particle reinforced 2080 aluminum alloy matrix composites. Materials Science and Engineering A 426, 314–322.

Christensen, R. M., Lo, K. H., 1979. Solutions for effective shear properties in three phase sphere and cylinder models. Journal of the Mechanics and Physics of Solids 27 (4), 315 – 330.

Christoffersen, J., 1983. Bonded granulates. Journal of the Mechanics and Physics of Solids 31 (1), 55 – 83.

Clausius, R., 1850. Ueber die bewegende Kraft der Wärme und die Gesetze, welche sich daraus für die Wärmelehre selbst ableiten lassen. Annalen der Physik, 368–397.

Clausius, R., 1865. Ueber verschiedene für die Anwendung bequeme Formen der Hauptgleichung der mechanischen Wärmetheorie. Vol. 125 of Annalen der Physik und Chemie. Johann Ambrosius Barth, pp. 353–401.

Coleman, B., Gurtin, M., 1967. Thermodynamics with internal state variables. The Journal of Chemical Physics 47, 597–613.

Cooper, D. W., Jul 1988. Random-sequential-packing simulations in three dimensions for spheres. Phys. Rev. A 38 (1), 522–524.

Cosserat, E., Cosserat, F., 1909. Théorie des corps deformables. Herman, Paris.

Decker, L., Jeulin, D., 2000. Simulation 3D de matériaux aléatoires poly-cristallins. Revue de Métallurgie - CIT/Science et Génie des Matériaux, 271–275.

Dederichs, P., Zeller, R., 1973. Variational treatment of the elastic constants of disordered materials. Zeitschrift für Physik (259), 103–116.

Dillard, T., N'guyen, F., Maire, E., Salvo, L., Forest, S., Bienvenu, Y., Bartout, J.-D., Croset, M., Dendievel, R., Cloetens, P., 2005. 3D quantitative image analysis of open-cell nickel foams under tension and compression loading using X-ray microtomography. Philosophical Magazine 85 (19), 2147 – 2175.

Du, Q., Faber, V., Gunzburger, M., 1999. Centroidal Voronoi Tessellations: Applications and Algorithms. SIAM Review 41 (4), 637–676.

Dvorak, G., Bahei-El-Din, Y., Wafa, A., 1994a. Implementation of the transformation field analysis. Computational Mechanics 14 (14), 201–228.

Dvorak, G., Bahei-El-Din, Y., Wafa, A., 1994b. The modeling of inelastic composite materials with the transformation field analysis. Modelling and Simulation in Material Science and Engineering 2 (2), 571–586.

Dvorak, G., Benveniste, Y., 1992. On transformation strains and uniform fields in multiphase elastic media. Proceedings of the Royal Society of London A 437 (437), 291–310.

Dvorak, G. J., Zhang, J., 2001. Transformation field analysis of damage evolution in composite materials. Journal of the Mechanics and Physics of Solids 49 (11), 2517 – 2541.

Döbrich, K., Rau, C., Krill III, C., 2004. Quantitative characterization of the three-dimensional microstructure of polycrstalline Al-Sn using X-Ray microtomography. Metallurgical and Materials Transactions A 35.

Eshelby, J.-D., 1957. The Determination of the Elastic Field of an Ellipsoidal Inclusion. Proceedings of the Royal Society of London A 241, 376–396.

Federov, F., 1968. Theory of Elastic Waves in Crystals. Plenum Press, New York.

Feyel, F., 2003. A multilevel finite element method (FE^2) to describe the response of highly non-linear structures using generalized continua. Computer Methods in Applied Mechanics and Engineering 192 (192), 3233–3244.

Fish, J., Shek, K., Pandheeradi, M., Shephard, M. S., 1997. Computational plasticity for composite structures based on mathematical homogenization: Theory and practice. Computer Methods in Applied Mechanics and Engineering 148 (1-2), 53 – 73.

Flaquer, J., Ríos, A., Martín-Meizoso, A., Nogales, S., Böhm, H., 2007. Effect of diamond shapes and associated thermal boundary resistance on thermal conductivity of diamond-based composites. Computational Materials Science 41 (2), 156–163.

Forest, S., Cailletaud, G., Sievert, R., 1997. A Cosserat theory for elastoviscoplastic single crystals at finite deformation. Archives of Mechanics 49 (4), 705 – 736.

Franciosi, P., Berbenni, S., 2007. Heterogeneous crystal and poly-crystal plasticity modeling from a transformation field analysis within a regularized Schmid law. Journal of the Mechanics and Physics of Solids 55 (11), 2265 – 2299.

Fritzen, F., Böhlke, T., 2009a. Homogenization of the physically nonlinear properties of three-dimensional metal matrix composites using the nonuniform transformation field analysis. Proceedings of the 17th International Conference on Composite Materials, Edinburgh, UK.

Fritzen, F., Böhlke, T., 2009b. Homogenization Of Three-Dimensional Micro-Heterogeneous Materials Using Nonuniform Transformation Fields. Proceedings of 7th EUROMECH Solid Mechanics Conference, Lisbon, Portugal.

Fritzen, F., Böhlke, T., 2010a. Influence of the type of boundary conditions on the numerical properties of unit cell problems. Technische Mechanik 30 (4), 354–363.

Fritzen, F., Böhlke, T., 2010b. Three-dimensional finite element implementation of the nonuniform transformation field analysis. International Journal for Numerical Methods in Engineering 84 (7), 803–829.

Fritzen, F., Böhlke, T., 2011a. Homogenized elasto-plastic response of high volume fraction metal ceramic composites based on nonuniform transformation fields.

Fritzen, F., Böhlke, T., 2011b. Nonuniform transformation field analysis of materials with morphological anisotropy. Composites Science and Technology 71, 433–442.

Fritzen, F., Böhlke, T., 2011c. Periodic three-dimensional mesh generation for particle reinforced composites with application to metal matrix composites. International Journal of Solids and Structures 48, 706–718.

Fritzen, F., Böhlke, T., Schnack, E., 2009. Periodic three-dimensional mesh generation for crystalline aggregates based on Voronoi tessellations. Computational Mechanics 43 (5), 701.

Ganapathysubramanian, S., Zabaras, N., 2004. Design across length scales: a reduced-order model of polycrystal plasticity for the control of microstructure-sensitive material properties. Computer Methods in Applied Mechanics and Engineering 193 (45-47), 5017 – 5034.

Ganesh, V., Chawla, N., 2005. Effect of particle orientation anisotropy on the tensile behavior of metal matrix composites: experiments and microstructure-based simulation. Materials Science and Engineering A 391, 342–353.

Garajeu, M., Michel, J. C., Suquet, P., 2000. A micromechanical approach of damage in viscoplastic materials by evolution in size, shape and distribution of voids. Computer Methods in Applied Mechanics and Engineering 183 (3-4), 223 – 246.

Garajeu, M., Suquet, P., 1997. Effective properties of porous ideally plastic or viscoplastic materials containing rigid particles. Journal of the Mechanics and Physics of Solids 45 (6), 873 – 902.

Garajeu, M., Suquet, P., 2007. On the influence of local fluctuations in volume fraction of constituents on the effective properties of nonlinear composites. Application to porous materials. Journal of the Mechanics and Physics of Solids 55 (4), 842 – 878.

Geers, M., Kouznetsova, V., Brekelmans, W., 2010. Multi-scale computational homogenization: Trends and challenges. Journal of Computational and Applied Mathematics 234 (7), 2175 – 2182, Fourth International Conference on Advanced Computational Methods in Engineering (ACOMEN 2008).

Germain, P., Nguyen, Q., Suquet, P., 1983. Continuum thermodynamics. Journal of Applied Mechanics 105, 1010–1020.

Gonzalez, R. C., Woods, R. E., 2002. Digital image processing, 2nd Edition. Prentice-Hall, Upper Saddle River, NJ.

Gurson, A., 1977. Continuum theory of ductile rupture by void nucleation and growth: Part I-yield criteria and flow rules for porous ductile media. Journal for Engineering Materials and Technology 99, 2–15.

Hackl, K., 1996. Generalized standard media and variational principles in classical and finite strain elastoplasticity. Journal of the Mechanics and Physics of Solids 45 (5), 667–688.

Halphen, N., Nguyen, Q., 1975. Sur les matériaux standards generalisés. Journal de Mécanique (14), 508–520.

Hashin, Z., Shtrikman, S., 1962a. On some variational principles in anisotropic and nonhomogeneous elasticity. Journal of the Mechanics and Physics of Solids 10, 335–342.

Hashin, Z., Shtrikman, S., 1962b. A variational approach to the theory of the elastic behaviour of polycrystals. Journal of the Mechanics and Physics of Solids 10, 343–352.

Hervé, E., Zaoui, A., 1993. n-Layered inclusion-based micromechanical modelling. International Journal of Engineering Science 31 (1), 1 – 10.

Hervé, E., Zaoui, A., 1995. Elastic behaviour of multiply coated fibre-reinforced composites. International Journal of Engineering Science 33 (10), 1419 – 1433.

Hill, R., 1952. The elastic behaviour of a crystalline aggregate. Proceedings of the Physical Society. Section A 65 (5), 349–354.

Hill, R., 1963. Elastic properties of reinforced solids: Some theoretical principles. Journal of the Mechanics and Physics of Solids 11 (5), 357 – 372.

Hill, R., 1965. A self-consistent mechanics of composite materials. Journal of Mechanics and Physics of Solids 13 (4), 213–222.

Holmes, C. C., Adams, N. M., 2002. A probabilistic nearest neighbour method for statistical pattern recognition. Journal of the Royal Statistical Society: Series B (Statistical Methodology) 64 (2), 295–306.

Jähne, B., 2005. Digitale Bildverarbeitung: mit Übungsaufgaben und CD-ROM, 6th Edition. Springer, Berlin.

Jöchen, K., Böhlke, T., Fritzen, F., 2010. Influence of the crystallographic and the morphological texture on the elastic properties of fcc crystal aggregates. Solid State Phenomena 160, 83–86.

Kamiński, M. M. (Ed.), 2005. Computational Mechanics of Composite Materials: Sensitivity, Randomness and Multiscale Behaviour. Engineering Materials and Processes. Springer-Verlag London Limited, London.

Kanit, T., May 2003. Notation de Volume Elementaire Répresentatif pour les Matériaux Héterogènes: Approche Statistique et Numérique. Ph.D. thesis, Centre des Matériaux P.M. FOURT de l'Ecole des Mines de Paris, B.P. 87, 91003 EVRY Cedex.

Kanit, T., Forest, S., Galliet, I., Mounoury, V., Jeulin, D., 2003. Determination of the size of the representative volume element for random composites: statistical and numerical approach. International Journal of Solids and Structures 40 (13-14), 3647 – 3679.

Kanit, T., N'Guyen, F., Forest, S., Jeulin, D., Reed, M., Singleton, S., 2006. Apparent and effective physical properties of heterogeneous materials: Representativity of samples of two materials from food industry. Computer Methods in Applied Mechanics and Engineering 195 (33-36), 3960 – 3982.

Kanoute, P., Boso, D., Chaboche, J., Schrefler, B., 2009. Multiscale Methods For Composites: A Review. Archives of Computational Methods in Engineering 16, 31–75.

Karhunen, K., 1946. Über lineare Methoden in der Wahrscheinlichkeitsrechnung. Vol. 37 of Al. Math. Phys. Ann. Acad. Sci. Fennicae.

Kruch, S., Carrère, N., Chaboche, J.-L., 2006. Fatigue damage analysis of unidirectional metal matrix composites. International Journal of Fatigue 28 (10), 1420 – 1425, The Third International Conference on Fatigue of Composites.

Kröner, E., 1958. Berechnung der elastischen Konstanten des Vielkristalls aus den Konstanten des Einkristalls. Zeitschrift für Physik 151 (4), 504–518.

Kröner, E., 1977. Bounds for the effective elastic properties of disordered materials. Journal of the Mechanics and Physics of Solids (25), 134–155.

Kumar, R. S., McDowell, D. L., 2009. Multifunctional design of two-dimensional cellular materials with tailored mesostructure. International Journal of Solids and Structures 46 (14-15), 2871 – 2885.

Kumar, S., Kurtz, S., 1994. Simulation of material microstructure using a 3d Voronoi tesselation: Calculation of effective thermal expansion coefficient of polycrystalline materials. Acta Metallurgica et Materialia 42 (12), 3917–3927.

Kumar, S., Kurtz, S. K., 1995. Monte-Carlo study of angular and edge length distributions in a three-dimensional Poisson-Voronoi tesselation. Materials Characterization 34 (1), 15 – 27.

Ladevèze, P., Passieux, J.-C., Néron, D., 2010. The LATIN multiscale computational method and the Proper Generalized Decomposition. Computer Methods in Applied Mechanics and Engineering 199 (21-22), 1287 – 1296, Multiscale Models and Mathematical Aspects in Solid and Fluid Mechanics.

Lautensack, C., Gietzsch, M., Godehardt, M., Schladitz, K., 2008. Modelling a ceramic foam using locally adaptable morphology. Journal of Microscopy 3 (230), 396–404.

Lebensohn, R. A., Brenner, R., Castelnau, O., Rollett, A. D., 2008. Orientation image-based micromechanical modelling of subgrain texture evolution in polycrystalline copper. Acta Materialia 56 (15), 3914 – 3926.

Leblond, J. B., Perrin, G., Suquet, P., 1994. Exact results and approximate models for porous viscoplastic solids. International Journal of Plasticity 10 (3), 213 – 235.

Li, J., 1999. On micromechanics approximation for the effective thermoelastic moduli of multi-phase composite materials. Mechanics of Materials 31 (2), 149–159.

Loève, M., 1963. Probability Theory. The University Series in Higher Mathematics. Van Nostrand, Princeton, NJ.

Lumley, J., 1967. The structure of inhomogeneous turbulent flow. Atmospheric Turbulence and Radio Wave Propagation.

Madi, K., Forest, S., Boussuge, M., Gailliègue, S., Lataste, E., Buffière, J.-Y., Bernard, D., Jeulin, D., 2007. Finite element simulations of the deformation of fused-cast refractories based on X-ray computed tomography. Computational Materials Science 39, 224–229.

Michel, J., Galvanetto, U., Suquet, P., 2002. Constitutive relations involving internal variables based on a micromechanical analysis. In: Drouot, R., Maugin, G., Sidoroff, F. (Eds.), Continuum Thermomechanics. Kluwer Academic Publishers, the Netherlands, pp. 301–312.

Michel, J., Suquet, P., 2003. Nonuniform transformation field analysis. International Journal of Solids and Structures 40, 6937–6955.

Michel, J., Suquet, P., 2004. Computational analysis of nonlinear composite structures using the nonuniform transformation field analysis. Computer Methods in Applied Mechanics and Engineering 193, 5477–5502.

Michel, J.-C., Suquet, P., 2009. Nonuniform transformation field analysis: a reduced model for multiscale nonlinear problems in solid mechanics. In: Galvanetto, U., Aliabadi, F. (Eds.), Multiscale Modelling in Solid Mechanics - Computational Approaches. Imperial College Press. Imperial College Press, London., pp. 159–206, ISBN: 978-1-84816-307-2. OR 1.

Miehe, C., 2002. Strain-driven homogenization of inelastic microstructures and composites based on an incremental variational formulation. Journal for Numerical Methods in Engineering 55, 1285–1322.

Miehe, C., Schotte, J., Schröder, J., 1999. Computational micro-macro transitions and overall moduli in the analysis of polycrystals at large strains. Computational Materials Science 16 (1-4), 372 – 382.

Mielke, A., 2004. Existence of minimizers in incremental elasto-plasticity with finite strains. SIAM Journal for Mathematical Analysis (63), 384–404.

Miserez, A., Rossoll, A., Mortensen, A., 2004. Investigation of crack-tip plasticity in high volume fraction particulate metal matrix composites. Engineering Fracture Mechanics 71, 2385–2406.

Mishnaevsky, L., 2007. Computational mesomechanics of composites: numerical analysis of the effect of microstructures of composites on their strength and damage resistance. Wiley, Chichester [u.a.].

Monchiet, V., Charkaluk, E., Kondo, D., 2007. An improvement of Gurson-type models of porous materials by using Eshelby-like trial velocity fields. Comptes Rendus Mécanique 335 (1), 32 – 41.

Mori, T., Tanaka, K., 1973. Average Stress in a Matrix and Average Elastic Energy of Materials with Misfitting Inclusions. Acta Metallurgica et Materialia 23, 571–574.

Moulinec, H., Suquet, P., 1998. A numerical method for computing the overall response of nonlinear composites with complex microstructure. Computer Methods in Applied Mechanics and Engineering, 69–94.

Murnaghan, F., 1962. The Unitary and Rotation Groups. Lecture on applied mathematics. Spartan Books, Washington.

Müller, I., 1985. Thermodynamics. Pitman, Boston.

Nemat-Nasser, S., Hori, M., 1999. Micromechanics: Overall properties of heterogeneous materials. Elsevier.

Nikolov, S., Petrov, M., Lymperakis, L., Friák, M., Sachs, C., Helge-Otto, F., Raabe, D., Neugebauer, J., 2010. Revealing the Design Principles of High-Performance Biological Composites Using Ab initio and Multiscale Simulations: The Example of Lobster Cuticle. Advanced Materials 22, 519–526.

Nogales, S., Böhm, H., 2008. Modeling of the thermal conductivity and thermomechanical behavior of diamond reinforced composites. International Journal of Engineering Science 46, 606–619.

Ohser, J., Mücklich, F., 2000. Statistical Analysis of Microstructures in Materials Science. Statistics in Practice. John Wiley & Sons.

Ortiz, M., Repetto, E., 1999. Nonconvex enery minimization and dislocation structures in ductile single crystals. Journal of the Mechanics and Physics of Solids 47, 397–462.

Osipov, N., Gourgues-Lorenzon, A. F., Marini, B., Mounoury, V., Nguyen, F., Cailletaud, G., 2008. FE modelling of bainitic steels using crystal plasticity. Philosophical Magazine 88, 3757–3777.

Ostoja-Starzewski, M., 2001. Scale and boundary conditions effects in elastic properties of random composites. Acta Mechanica 148, 63–78.

Paufler, P., Schulze, G., 1978. Physikalische Grundlagen mechanischer Festkoerpereigenschaften. Vieweg, Braunschweig.

Pierard, O., Firebel, C., Doghri, I., 2004. Mean-field homogenization of multi-phase thermoelastic composites: a general framework and its validation. Composites Science and Technology 64, 1587–1603.

Ponte-Castañeda, P., 1991. The effective mechanical properties of nonlinear isotropic composites. Journal of the Mechanics and Physics of Solids 39 (1), 45 – 71.

Qu, J., Cherkaoui, M., 2006. Fundamentals of micromechanics of solids. Wiley, Hoboken [u.a.].

Ragnemalm, I., 1992. Fast erosion and dilation by contour processing and thresholding of distance maps. Pattern Recognition Letters 13 (3), 161 – 166.

Renard, J., Marmonier, M.-F., 1987. Etude de l'initiation de l'endommagement dans la matrice d'un matériau composite par une méthode d'homogénisation. Aerospace Science and Technology 6, 37–51.

Reuss, A., 1929. Berechnung der Fließgrenze von Mischkristallen auf Grund der Plastizitätsbedingung für Einkristalle. Zeitschrift für Angewandte Mathematik und Mechanik 9 (1), 49–58.

Rosen, B. W., Hashin, Z., 1970. Effective thermal expansion coefficients and specific heats of composite materials. International Journal of Engineering Science 8 (2), 157 – 173.

Roussette, S., Michel, J., Suquet, P., 2009. Nonuniform transformation field analysis of elastic-viscoplastic composites. Composites Science and Technology 69, 22–27.

Rychlewski, J., 1995. Unconventional approach to linear elasticity. Arch. Mech. 47 (2), 149–171.

Ryckelynck, D., 2009. Hyper reduction of mechanical models involving internal variables. International Journal of Solids and Structures 77 (1), 75–89.

Ryckelynck, D., Benziane, D., 2010. Multi-level a priori hyper reduction of mechanical models involving internal variables. Computer Methods in Applied Mechanics and Engineering 199, 1134–1142.

Ryckelynck, D., Benziane, D., Musienko, A., Cailletaud, G., 2010. Toward green mechanical simulations in materials science: hyper-reduction of a polycrystal plasticity model. European Journal of Computational Mechanics 19 (4), 365–288.

Schöberl, J., 1997. NETGEN An advancing front 2D/3D-mesh generator based on abstract rules. Computing and Visualization in Science 1 (1), 41–52.

Shewchuk, J., May 1996. Engineering a 2D Quality Mesh Generator and Delaunay Triangulator. In: Lin, M., Manocha, D. (Eds.), Applied Computational Geometry: Towards Geometric Engineering. Springer-Verlag, pp. 203–222.

Si, H., Gaertner, K., Sep. 2005. Meshing Piecewise Linear Complexes by Constrained Delaunay Tetrahedralizations. In: Proceedings of the 14th International Meshing Roundtable. pp. 147–163.

Simmons, G., Wang, H., 1971. Single Crystal Elastic Constants and Calculated Aggregate Properties: A Handbook. The M.I.T. Press.

Smit, R. J. M., Brekelmans, W. A. M., Meijer, H. E. H., 1998. Prediction of the mechanical behavior of nonlinear heterogeneous systems by multi-level finite element modeling. Computer Methods in Applied Mechanics and Engineering 155 (1-2), 181 – 192.

Stainier, L., Ortiz, M., 2010. Study and validation of a variational theory of thermomechanical coupling in finite visco-plasticity. International Journal of Solids and Structures (47), 705–715.

Sun, Y., Wang, D., 1989. Acta Mechanica Sinica, 237–243.

Suquet, P., 1985a. Elements of homogenization for inelastic solid mechanics. In: Snachez-Palencia, E., Zaoui, A. (Eds.), Homogenization Techniques for Composite Media. Vol. 272 of Lecture Notes in Physics. Springer Verlag.

Suquet, P., 1985b. Local and global aspects in the mathematical theory of plasticity. pp. 279–310.

Suresh, S. (Ed.), 2002. Fundamentals of metal matrix composites. Knovel, Norwich, NY.

Torquato, S., 2002. Random heterogeneous materials: microstructure and macroscopic properties. Springer, New York [u.a.].

Truesdell, C., Noll, W., 1965. The Non-Linear Field Theories of Mechanics. Vol. III/3 of Handbuch der Physik. Springer.

Tvergaard, V., 1981. Influence of voids on shear band instabilities under plane strain conditions. International Journal of Fracture 17 (4), 389–407.

Voigt, W., 1910. Lehrbuch der Kristallphysik. Teubner, Berlin.

von Mises, R., 1928. Mechanik der Formänderung von Kristallen. Zeitschrift für Angewandte Mathematik und Mechanik, 161–185.

Walpole, L., 1966a. On bounds for the overall elastic moduli of inhomogeneous systems–I. Journal of the Mechanics and Physics of Solids 14 (3), 151 – 162.

Walpole, L., 1966b. On bounds for the overall elastic moduli of inhomogeneous systems–II. Journal of the Mechanics and Physics of Solids 14 (5), 289 – 301.

Williams, W., Smith, C., 1952. A study of grain shape in an aluminium alloy and other applications of stereoscopic microradiology. Transactions, American Institute of Mining Engineers 194, 755–765.

Willis, J., 1977. Bounds and self-consistent estimates for the overall properties of anisotropic composites. Journal of the Mechanics and Physics of Solids 25, 185–202.

Yang, Q., Stainier, L., Ortiz, M., 2006. A variational formulation of the coupled thermomechanical boundary-value problem for general dissipative solids. Journal of the Mechanics and Physics of Solids 54, 401–424.

Zeller, R., Dederichs, P., 1973. Elastic Constants of Polycrystals. Physica Status Solidi (b) 55 (2), 831–842.

Zhodi, T., Wriggers, P., 2005. Introduction to computational micromechanics. Springer.

Zienkiewicz, O., Taylor, R., Zhu, J. Z., 2006. Finite element method, 6th Edition. Butterworth-Heinemann.